Stuart Yarno

T0153527

Arduino

In easy steps is an imprint of In Easy Steps Limited
16 Hamilton Terrace · Holly Walk · Leamington Spa
Warwickshire · United Kingdom · CV32 4LY
www.ineasysteps.com

Notice of Liability
Every effort has been made to ensure that this book contains accurate
and current information. However, In Easy Steps Limited and the
author shall not be liable for any loss or damage suffered by readers
as a result of any information contained herein.

Trademarks
All trademarks are acknowledged as belonging to their respective
companies.

In Easy Steps Limited supports The Forest Stewardship Council (FSC),
the leading international forest certification organisation. All our titles
that are printed on Greenpeace approved FSC certified paper carry the
FSC logo.

MIX
Paper from
responsible sources
FSC® C020837

Printed and bound in the United Kingdom

ISBN 978-1-84078-633-0

Contents

8   Programming Arduino   111

9   Sketches   137

10   Troubleshooting & Debugging   165

11 Arduino Projects 175

Index 187

1 Understanding Arduino

This chapter is an introduction to the subject of Arduino. We see just what Arduino is, what can be done with it and the advantages it offers over competing platforms.

What is Arduino?

An Arduino is a small circuit board, which contains either an 8-bit or a 32-bit microcontroller, plus a handful of other components. Recent models, such as the Uno, also offer a USB interface, and a number of analog input pins as well as a number of digital input/output pins.

The concept behind the development of Arduino is to simplify the construction of interactive objects or environments and make them more accessible. To this end, it has been designed to be inexpensive and straightforward, thus providing a way for hobbyists, students, and professionals to create devices and projects that interact with their environment with sensors and actuators.

Typical examples of Arduino projects include simple robots, security systems and motion detectors. There are many more.

Arduino is about more than just hardware though. The microcontroller needs to be programmed and this introduces a software element in the form of an integrated development environment (IDE) that runs on personal computers. With it, users write programs (known as sketches) using the C or C++ programming languages.

The Arduino's microcontroller comes with a boot loader that makes it much easier to upload programs to the board's flash memory. The Arduino's competitors, in comparison, typically require the use of an external programmer. In keeping with the ethos behind Arduino, this keeps the programming side of things as straightforward as possible by allowing it to be done with a personal computer.

Because the connectors on Arduino boards are all standard types, it is possible to extend the basic Arduino by incorporating pre-built circuit boards, known as shields. Due to the design of the connector blocks, these can be stacked on top of one another, making it possible to create large and complex projects.

A number of Arduino boards are available. These include the Uno, the Duemilanove, the Diecimila, and the Mega. These are all designed for use with specific types of project and so come with differing specifications. Of all of them, the Uno is the most popular and can be used for the widest range of projects.

Arduino began in Italy at the Interaction Design Institute Ivera. This is a school of design education that focuses on interaction with digital devices and systems.

Arduino is the modern-day equivalent of those old electronic kits from yesteryear sold by companies such as Radio Shack and Heath

As a way of learning the basics of both electronics in general and programming, Arduino is an ideal tool.

Why Arduino?

The Arduino isn't the only low-cost computer on the market. There are a number of alternatives, all of which provide much the same capabilities. These include:

- Raspberry Pi
- CubieBoard
- Gooseberry
- APC Rock
- OLinuXino
- Hackberry A10

However, Arduino does have the following advantages:

Cost – Arduino boards are not as expensive as other microcontroller platforms. For many people, this is an important factor.

Cross-platform – Arduino software runs on Windows, Mac, and Linux operating systems. This is in contrast to most other microcontroller systems, which are limited to just Windows.

Simplicity & Flexibility – the Arduino programming environment is simple enough for beginners to quickly grasp, yet provides the flexibility that enables more complex projects to be tackled.

As an educational tool, it's conveniently based on the common open source Processing programming environment, so many people will already be familiar with it.

Open source software – Arduino software is open source and freely available for download from a number of websites. For those who don't have the ability to write their own programs or those who simply cannot be bothered, a huge range is available.

Open source hardware – the Arduino is based on the ATMEGA8 and ATMEGA168 microcontrollers, the plans for which are freely available under a Creative Commons license. Thus users can make their own version of the module to improve it or to increase its capabilities.

While we discuss Arduino in this book, it may be that one of the other low-cost computers is better suited to your requirements.

Arduino does offer advantages over the competition.

Which Arduino?

Arduino boards are not all the same – there are a number of types, each designed for different requirements. So before parting with the cash, it may pay you to take a look at what is available. Below, we look at the three most popular boards:

Arduino Uno

The most popular Arduino board of all is the Uno. This offers a range of features that make it a good all-purpose board.

It incorporates the ATmega328 chip as the controller and can be powered directly from USB, battery or an AC to DC adapter. The board operates at a voltage of 5 volts.

The board also offers 14 digital input/ output pins, six of which can be used as pulse width modulation (PWM) outputs. In addition, it has six analog inputs, plus RX/TX (serial data) pins.

Memory provision is 32KB of flash memory, 2KB of static random access memory (SRAM), and 1KB of Electrically Erasable Programmable Read-Only Memory (EEPROM).

Variations of the Uno include:

The Leonardo – this is available both with and without headers, and offers a micro USB port.

The Ethernet – this comes with a RJ45 Ethernet socket rather than a USB port. It also offers a microSD card reader.

Pros of the Arduino Uno include:

- Simplicity
- Low cost
- Plug-and-play
- Plenty of resources available, e.g. tutorials, sample code, etc.
- Numerous extras available, e.g. shields and libraries

...cont'd

Cons include:

- Number of input/output pins is limited
- Small amount of memory provided can be restrictive
- 8-bit microcontroller

Arduino Mega 2560

This board is the Uno's big brother and is much the same, apart from being bigger. The increase in surface area allows it to offer 70 I/O pins (the Uno has just 14). Of these, 16 are analog inputs with the other 54 being digital. 15 of the digital pins can handle pulse width modulation (PWM). Also included are four RX/TX serial ports.

The Mega is designed for more advanced users.

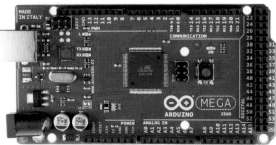

With regard to the controller, this is the ATmega2560 chip, which operates at 5 volts.

The Mega provides some four times more memory than the Uno. Specifically, 256KB of flash memory, 8KB of SRAM and 4KB of EEPROM.

Variations of the Mega include:

The Due – this is based on a 32-bit ARM core microcontroller. Being much faster than the Uno, it is designed for use with more demanding applications. To this end, it is also furnished with more memory – 512KB of flash and 96KB of SRAM. Unlike other Arduino boards, the Due runs at 3.3 volts.

The ADK – this board is intended for use with Android cell phones.

If you are building a project for use with an Android cell phone, the Arduino ADK is the board to use.

Pros of the Arduino Mega 2560 include:

- Plenty of input and output pins
- Good amount of memory
- Plenty of resources available, e.g. tutorials, sample code, etc.
- Provides scope for bigger projects than does the Uno

...cont'd

Cons include:

- More expensive than the Uno (typically twice the price)
- Resources not as plentiful as for the Uno

Arduino Pro

As might be deduced from the name, the Arduino Pro is designed for professional use. The board is based on the ATmega168 or ATmega328 microcontrollers.

Of all the Arduino boards, the Pro offers the most options and flexibility.

The Pro comes in both 3.3 volt and 5 volt versions. It has 14 digital input/output pins (six of which can be used as PWM outputs), and six analog inputs.

It also has holes for mounting a power jack, an ICSP header, and pin headers. A 6-pin header can be connected to an FTDI cable to provide USB power and communication to the board.

The Arduino Pro is designed for use in semi-permanent installations. The board doesn't provide pre-mounted headers thus allowing the user to employ whatever types of connectors are required by the project at hand.

Pros of the Arduino Pro include:

- Well suited for use in embedded projects
- Provides good flexibility when designing projects
- Soldered joints results in a higher level of reliability

Cons include:

- More expensive than the Uno
- Joints/connections need to be soldered

Connections in the Arduino Pro need to be soldered by hand.

12

What Can You Do With It?

Now that you have some insight into what Arduino is, why you should use it in preference to competing platforms, and which version to use; you may want to consider just what can be done with it.

The first thing to be aware of is that because it comes with a microcontroller, an Arduino board can be used as the brains behind virtually any electronics project. This adds a huge amount of versatility over the simple electronic kits of yesteryear.

By connecting a range of switches and sensors, Arduino can "sense" its surroundings; this includes light, sound, temperature, motion, pressure, etc. It can take this information and then with the aid of motors and other actuators, use it to interact with what's around it. As a random and very basic example, you could use Arduino to automatically control the lights in your house.

Devices that you can control with Arduino include switches, LEDs, motors, loudspeakers, GPS units, cameras, the Internet, and even smartphones and TVs.

Arduino can operate as an independent unit, it can be connected to a computer, or be connected to other Arduinos, or other electronic devices. In fact, with a bit of ingenuity pretty much anything can be connected to, and controlled by, Arduino.

Some Arduino projects are simply for fun. For example, a flashing LED cube, a tree climbing robot, a laser harp. Most, though, have a practical use. Typical examples are security systems, automated plant watering and smartphone garage door controllers. Arduino also provides a simple and inexpensive platform for building prototypes that enables ideas to be evaluated and turned into reality.

However, there are applications for which Arduino is inherently unsuited. Because it has a limited amount of memory and slow processing capabilities, Arduino cannot be used for anything that requires serious processing power. This rules out video and audio processing, recording and output.

It also has a high power consumption, which means that battery powered applications can use up batteries at an alarming rate.

While you can do many things with a basic Arduino kit, complex projects will require extra parts.

Arduino is used by artists, designers, and hobbyists.

Basic Principles

Before we get down to the nitty gritty of Arduino, we'll take a look at the basics. This may help you decide whether or not Arduino is for you.

Hardware

An Arduino project is comprised of both hardware and software elements. We'll start with the former.

Breadboard – in your Arduino kit, there will be a blank board known as a breadboard, as shown below:

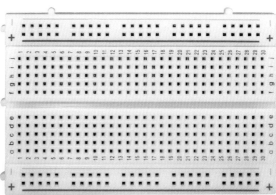

Hot tip

Soldering skills are not essential for Arduino.

You will use this to build circuits. These are constructed by pushing the leads of electronic components such as resistors and capacitors into the breadboard's holes and then connecting them up with jumpers – all you need to know is which holes to push them into.

Note that there is no need to solder components in place – a big plus for those who have never soldered before or simply don't want the bother of it.

Breadboards offer two advantages: The first is that they enable circuits to be built quickly and easily. The second is that they make it very easy to rearrange components in order to correct mistakes or simply experiment.

Components – your Arduino kit includes a number of components, such as resistors, capacitors, switches, motors, etc. – see page 80 for a full list. These are the building blocks of your projects.

Shields – the parts supplied in your kit are sufficient to build very basic projects. Once you get past these though, you will discover the need for more advanced circuits or, indeed, several of them, to build projects.

If you have the ability there is nothing to stop you building these yourself with extension circuits. If you don't though, or simply don't want the bother, you can buy the required circuitry in the form of pre-built printed circuit boards. These are known as shields

Arduino Ethernet Shield

and they are readily available from a number of retail outlets.

The Microcontroller – having put together the various elements that comprise a project, you then need to give the Arduino the instructions necessary to make the project work, i.e. you need to program it.

The part of the Arduino board that interprets the instructions, or program, is the microcontroller.

However, before you can get started with programming, you will first need to set up the Arduino's software.

Software

This is not supplied with the board. You have to go to the Arduino website at **www.arduino.cc/en/main/software** and download it from there. The software is known as the integrated development environment (IDE).

There are IDE versions for Windows, Mac OS X, and Linux so you will be able to use Arduino with whatever operating system is on your PC.

Simply follow the prompts to install the software.

Don't forget

If you don't have either the time or skills to build circuits yourself, you can buy ready-made shields.

Hot tip

To program your projects you will need the use of a computer.

15

...cont'd

Programming – Having installed the IDE, you will now be able to write the necessary code for your project. This is done within the IDE on your computer, which provides a code editing window as shown below. The programs you write are known as sketches and can be manipulated in the same way as any other data, e.g. saved to the PC, uploaded to the Internet, etc.

C Programming in easy steps and C++ Programming in easy steps are available to learn how to program in these languages.

With the sketch written, connect the Arduino board to the computer with the supplied USB cable and upload the code to the microcontroller. At this point your project is complete – all that remains is to see if it works.

Required Skills

The Arduino system is designed to be as straightforward as possible. However, that doesn't necessarily mean you can step straight in and get immediate results. Arduino projects involve more than just the Arduino itself. You will need other skills:

One is programming – all Arduino projects need to be programmed using the C and C++ programming languages. Another is electronics. All but the most simple of projects require external circuits to be constructed. This requires knowledge of electronic components, circuit design and construction.

Other skills you are likely to need include design, mechanics, fabrication and computing, to name just some. Obviously, this depends enormously on the scope of your projects.

For those of you who don't have the requisite knowledge, the Internet is an enormous resource for all things to do with Arduino.

2 The Arduino Kit Bag

In this chapter, we look at the contents of your Arduino kit and explain what the various items are for.

The Arduino Uno Board

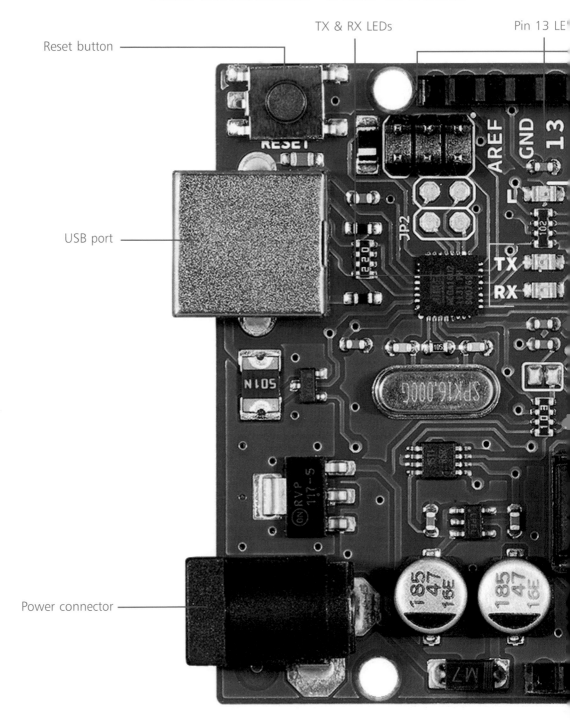

Reset button

TX & RX LEDs

Pin 13 LE

USB port

Power connector

Digital pins

~9 8 7 ~6 ~5 4 ~3 2 TX→1 RX←0

DIGITAL (PWM~)

UNO

ARDUINO

ON

ICSP

WWW.ARDUINO.CC – MADE IN ITALY

ATMEGA328P-PU

ATMEL

POWER

5V GND GND Vin

ANALOG IN

A0 A1 A2 A3 A4 A5

Power LED

Microcontroller

Power pins

Analog pins

Important Board Parts

In Chapter One, we described Arduino in general terms; we will now look at the hardware aspect in more detail. We'll start with the Arduino board itself.

On pages 18-19 we see the layout of the most popular board, the Uno, and its main components. These are:

Microcontroller

This integrated circuit, or IC, is the brains of the board and provides its computational power. Different versions are used in different Arduino boards and the one employed in the Uno is the Atmel ATmega328.

Different Arduino boards have different capabilities. This is largely down to the microcontroller.

Essentially, it is a tiny computer in its own right. Within the IC is a controller that executes instructions, and several different types of memory that store data and instructions. There are also a number of individual paths for the sending and receiving of data.

Header Sockets

The microcontroller is connected to a number of sockets, known as header sockets or pins. These are arranged around the top and bottom edges of the board, and comprise three types: digital pins, analog pins and power pins.

Extension boards (known as shields) also have header sockets and enable boards to be stacked on top of each other when creating complex projects.

Their purpose is to enable additional circuitry, such as breadboards and shields, to be connected to the Arduino board quickly and easily; and to provide power to the board.

Digital pins – These run across the top of the board. The pins numbered 0 to 13 send and receive digital signals. Pins 0 and 1 also act as a serial port, which is used to send and receive data to other devices, e.g. a computer, via the USB connector.

Note that by digital we mean a signal that has one of two possible states: on or off. With the Arduino Uno, this means a voltage of either 0 or 5 with nothing in between.

Analog in pins – At the bottom of the board are two more rows of pins. Labeled A0 to A5 are six pins that provide analog inputs. Pins A4 and A5 can also be used to send and receive data from other devices.

Note that by analog we mean a signal that varies in strength between two fixed values. With the Auduino Uno, this means a voltage that varies in strength anywhere in the 0 to 5 volt range.

Analog out pins – these are located at the top of the board and are labeled with a tilde (~). These pins use a technique known as pulse width modulation (PWM), which essentially simulates an analog signal using digital pins. You will notice that the ~ appears alongside pins 3, 5, 6, 9, 10 and 11 – indicating that these can all handle PWM.

The PWM signal is used to control devices such as motors that require an analog signal.

Power pins – At the the bottom-left of the board are the eight power pins. These are used to provide and distribute power wherever it is required.

The Vin pin (voltage in) can act as a external voltage source input to the Arduino board, or as a voltage output source to power external components.

Pin2 is the IOREF pin and is used to tell any shield attached to the board what voltage the host Arduino board is running at.

The two GND pins enable circuits to be completed. There is also another GND pin at pin 13 at the top of the board. These pins are all linked and share the same common ground.

The 5V pin is used to supply 5 volt power to the board and the 3.3V pin to supply 3.3 volt power.

USB Socket
At the top-left side of the board is the USB socket. With a USB cable, you can connect the board to a computer so that sketches can be uploaded to the microcontroller.

The difference between digital and analog signals is important. Digital signals are fixed at either 0V or 5V. Analog signals vary anywhere between 0V and 5V.

By utilizing the Vin pin, you can use your Arduino board to power external boards.

Note that you can also power your Arduino board via USB.

Beware

An external power supply of more than 12 volts may damage your board.

22

...cont'd

The socket accepts the type B USB connector. This is supplied as part of the kit and it is the same cable as commonly used by printers and scanners, should you ever misplace it.

External Power Connector

At the lower-left side of the board is the external power connector or jack. External (non-USB) power can come either from an AC to DC adapter or a battery.

The adapter can be connected by plugging a 2.1mm center-positive plug into the board's power connector. Leads from a battery can be inserted in the GND and Vin pin headers of the power connector. If you attempt this, you need to be careful as getting the connections the wrong way round will almost certainly trash the board.

The Arduino will function on an external supply of 6 to 20 volts. If supplied with less than 7 volts, however, the 5 volt pin may supply less than five volts and the board might be unstable as a result. If supplied with more than 12 volts, the voltage regulator may overheat and damage the board. The recommended range is thus 7 to 12 volts.

LEDs

Your Arduino has four LEDs. At the right of the board is the power LED which lights up when the board is receiving power.

The TX and RX LEDs, respectively, indicate that the board is transmitting or receiving data between the Arduino and external devices via either the serial port or the USB port.

The L LED (just above the TX LED) is connected to digital pin 13 and when lit indicates that the board is functioning correctly.

Reset Button

As with any computerized device, things don't always go according to plan. For whatever reason, the task at hand may refuse to work. To resolve unexpected issues, the board provides a reset button. This is located at the top-left hand corner.

Pressing it down will reset the currently running sketch, and holding it down will stop it completely. Note that you can also perform these actions from the Arduino IDE on your computer.

Breadboards

We took a very brief look at breadboards on page 14. Here, we'll see why they play such an important role in Arduino projects.

The core principle of Arduino is that of prototyping, i.e. the exploration of ideas. It's important to have a method of doing this that is quick and also inexpensive; breadboards provide this method.

They do it by providing a base for the building of prototype circuits that is temporary rather than permanent due to not having to solder components in place.

So having built and tested a circuit, and confirmed that it works, you can then make it permanent by soldering. If, on the other hand, it doesn't work, simply pull the components out of the breadboard and start again.

Terminal Strips

The construction of a breadboard is basically a plastic case with columns of holes on the surface. Each column is labeled with a letter, i.e. a, b, c, d, etc. Running down the board are numbered rows, and on the underside of the board are conductive terminal strips made of copper.

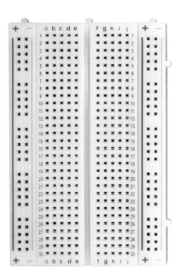 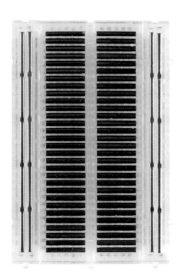

This is demonstrated in the image above. On the left is the topside of a breadboard showing the rows and columns; on the right is the underside showing the terminal strips.

Beware

Solderless breadboards are only suitable for low frequency circuits. They are also unsuitable for complex circuits due to the amount of wiring required.

Hot tip

Each row in a board is labeled with a number and each column with a letter. These are there as a guide as things can get complicated quite quickly when building a circuit.

...cont'd

Just below the surface of each hole is a metal clip. When the lead of a component is pushed into a hole, the clip grabs it and holds it in place. Once inserted, the component will be electrically connected to anything else placed in that row.

Trench

Down the center of the breadboard is a gap, known as a trench. This splits each row into two electrically separate sections.

Hot tip

The trench is also useful for fitting integrated circuits where the legs on each side need to be kept electrically separate.

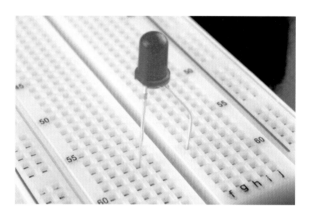

This enables components to be placed in the board without their pins being shorted together as we see in the image above.

Power Rails

Running vertically down either side of the breadboard are power rails. These rails are made of copper strips and are identical to the terminal strips.

When building circuits, power tends to be required in different places. The power rails provide this access. They are labeled with a + and a - and also have a red and blue or black stripe that indicates the positive and negative sides.

Don't forget

Power rails run the entire length of the breadboard thus ensuring power is always available to all parts of a circuit.

In larger breadboards, you may find one half of the board's power rails is isolated from the other half so there is a top and bottom half on each side.

This is useful in situations where two different voltages are required to power a circuit. You need to be careful though as it's easy to connect a part to the wrong rail and maybe burn it out.

24

Jump Wires

Jump wires are another essential part of any electronic project. They are short pieces of insulated wire with exposed ends, and are used to connect components in the breadboard.

Your Arduino kit starts you off with a few jump wires but it probably won't be long before you need more. They can be bought in packs of assorted colors and lengths or cut to length from a reel.

Of the two, we prefer the packs. The different colors are helpful for identification purposes, plus they come with rigid points which makes it easier to connect them in a breadboard.

However, there are advantages in cutting them yourself. First, you can cut the exact length needed which can help enormously to keep the circuit tidy. Second, they are less expensive.

Components

In your Arduino kit, you will find a box containing a number of electronic components.

It's not strictly essential that you learn what these do and how they do it but, if you don't, you will never be able to design and build your own projects – you will only ever be

able to build projects by copying other people's efforts.

As there is little point in an approach such as this – the whole idea of Arduino is one of innovation and experimentation – we recommend you take the time to learn about these components.

Chapter 6 is devoted to this subject and, while it can be no more than a primer given the space available, it will give you a basic grounding in the components used in electronic circuits.

Hot tip

Breadboards can get crowded quickly and so it is not always possible to position components just where you'd like them. Jump wires provide the solution.

Hot tip

Another type of jump wire is supplied in pre-cut lengths and already angled. Being rigid, they lie flat against the board thus helping to eliminate what can be a confusing tangle of wires.

Don't forget

If you want to design your own projects, you will need to learn the basics of electronics and the components that make it work.

Base Plates

Many Arduino projects comprise both the Arduino board and the breadboard. As these are two separate entities linked only by connecting wires, it is necessary to mount them on something so they are effectively a single unit. This is achieved with the aid of a base plate or board.

In your Arduino kit, you will find a pre-cut self-assembly wooden base that will do the job, as shown on the right. Assembly instructions are in the Arduino manual.

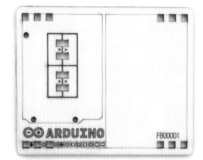

The Arduino board is fixed to the base plate with the supplied nuts and bolts, while the breadboard has a self-adhesive backing that allows it to be fixed alongside the Arduino board.

However, it has to be said that much better base plates are available. Made of acrylic and offering mounting holes for the most popular Arduino boards, these items cost just a few dollars and are well worth the investment.

To make life easier, we recommend you invest in a good quality base plate.

A good example is shown above. Boards of this type come with rubber feet that help prevent the assembly slipping about on the work surface.

3 Arduino Software

In this chapter, we look at the software side of Arduino. We show how to set it up and then explain what the various options do.

Install Arduino on Windows

Software is not supplied with the Arduino – you need to download it to your computer from the Arduino website.

We refer to Windows 8 here but the same applies to Windows 8.1.

1 Go to **http://arduino.cc/en/main/software**

You will see download links for several versions of Arduino. The one you want is version 1.0.5.

2 Under Arduino 1.0.5, click the Windows Installer link. The download will commence, and when completed the installation routine will begin automatically

Most hardware devices supply the software on a disc. With Arduino, the software has to be downloaded from the Arduino website.

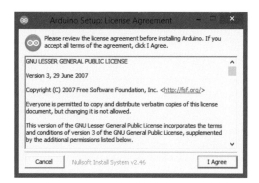

3 Click **I Agree** in the license agreement window

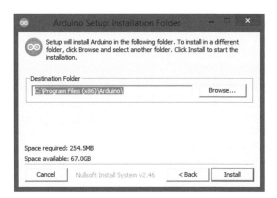

Here we see Arduino being installed on Windows 8. The procedure is much the same for other editions of Windows.

4 By default, Arduino will be installed to C:\Program Files (x86)\Arduino. If you want to install it to a different location, click the **Browse** button and select the required folder. Then click **Install**

5 When prompted for the USB driver, click Install. When the installation is complete you will see an Arduino icon on the Desktop. Click it to open the Arduino IDE

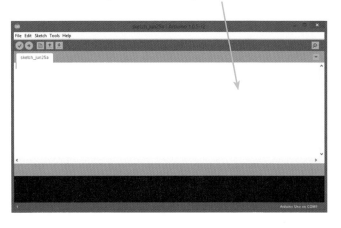

Install Arduino on Mac OS X

1 Go to **http://arduino.cc/en/main/software**

These instructions show Arduino being installed on Mac OS X 10.8 "Mountain Lion".

2 Under Arduino 1.0.5, click the Mac OS X link to begin the download. When complete go to your Downloads folder where you will see the Arduino icon

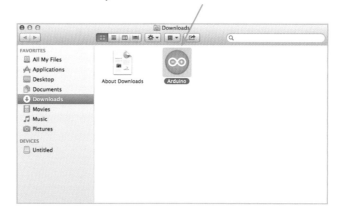

Don't forget

The Dock provides a handy location for frequently used applications.

3 Drag the icon to the Applications folder. Alternatively, you can drag it to the Dock from where it is quickly accessible

...cont'd

4 Click the Arduino icon to open the program. You will see a warning message saying Arduino is a program downloaded from the Internet and do you want to open it? It is quite safe to do so, so click **Open**

5 The Arduino IDE opens. You can now begin to write code with which to program the microcontroller

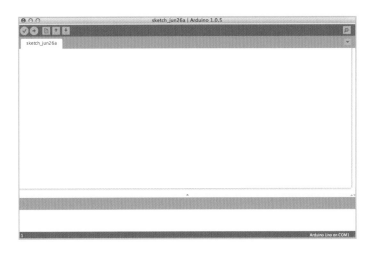

Install Arduino on Linux

Installing Arduino on Linux requires a slightly different approach than for Windows and Mac OS X. There are many different versions of Linux and the following example shows how to install it on Ubuntu, which is one of the most popular.

Other popular versions of Linux include Debian, Mint and Fedora.

1 On the Ubuntu desktop, click the Ubuntu Software Center icon. When it opens, type Arduino in the search box at the top-right of the window as shown below

2 The search utility will automatically locate the correct version of Arduino. When you see it in the search results list, click on it to reveal the Install button at the right

3 When the installation is complete, go to the desktop and click the "Search your computer and online sources" icon at the top-left. The first item in the Applications section will be Arduino as shown above. Click the icon to open the program

4 A window will open saying you need to be added to the dialout group before you will be able to upload code to the Arduino microcontroller. Click the **Add** button

The installation instructions shown here are for the Ubuntu version of Linux.

5 Authenticate the above action by entering your Ubuntu password in the next window and then clicking **Authenticate**

6 An Arduino icon will now appear on the sidebar. Click on it to open the Arduino IDE as shown above

Setting Up Arduino

Setting up Arduino is basically the process of connecting the board to the computer and then installing the drivers. Problems can occur at this stage but they tend to happen more with computers running older operating systems.

The first thing you have to do is connect the board to your computer with the USB cable provided in the kit. Once you have done this, the board's green power LED should light up indicating it is receiving power.

What happens next depends on what operating system you are using.

Windows

Recent Windows operating systems work well with Arduino. If you are running Windows 8 in particular, setting up will be a breeze. Early operating systems (Windows XP and earlier) may prove to be troublesome.

Connect the board to the computer and after a few moments the Device Setup window will appear. Note that in this example, we are setting up Arduino on Windows 8.

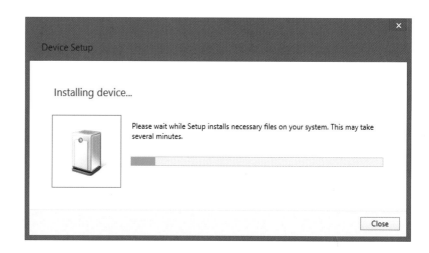

Windows 8 will automatically locate and install the Arduino driver – you don't need to do anything at all.

The installation only takes a few moments and then the window closes.

Arduino will work with virtually any operating system. However, the older it is, the more problematic it will be to set Arduino up.

With Windows operating systems, Arduino will always use the COM3 port.

34

To check that the driver is installed correctly, go to the Device Manager in the Control Panel. Click the arrow next to Ports and you should see that Arduino is configured to use COM3, as shown below:

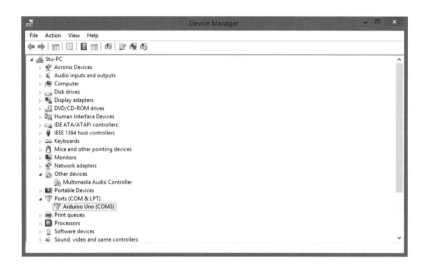

Mac OS X

The Arduino setting up procedure for Mac OS X Lion, Mountain Lion, Leopard and Snow Leopard is the same and should be straightforward. Earlier editions may present problems.

1 Connect the board to the computer with the USB cable. A dialog box will open as shown below:

2 Click **Network Preferences** and in the next window, click **Apply**. If you look at the left-hand side of this window, you will notice that Arduino is displayed as being "Not Configured". Take no notice of this – Arduino is correctly set up and ready for use.

For other editions or if you experience problems setting up take a look at http://forum.arduino.cc/

The "Not Configured" message is erroneous – ignore it.

...cont'd

Linux

With modern Linux operating systems, such as Ubuntu 14.04, Arduino does not require any setting up at all.

Simply connect the board with the USB cable, open the Arduino software and you are ready to go.

However, with older versions problems are likely to occur. We suggest consulting one of the many websites devoted to this issue.

For example, go to **http://playground.arduino.cc/learning/linux** Here you will find detailed instructions for installing Arduino on all versions of Linux. Just select yours from the list provided.

If you are having problems setting up Arduino on Linux, we recommend you take a look at http://playground.arduino.cc/learning/linux

Check It's Working

Having set up Arduino, it will be as well to make sure the Arduino board is actually communicating with the computer before you go any further. Do this as follows:

1 Connect the board to the PC with the USB cable

2 Open the Arduino IDE on your computer

3 From the File menu, select **Examples > Basics > Blink**

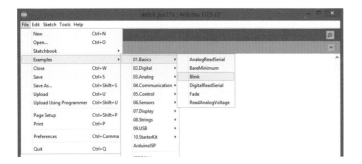

4 This loads the "Blink" sketch, into the code editing window. Click the **Upload** icon at the top-left

Uploading a sketch to the microcontroller on the Arduino board and seeing the resultant change in the board's behavior shows that everything is working as it should.

You should see two things: First, the Done uploading message towards the bottom-left of the screen. Second, the pin 13 LED on the board will now be blinking on and off. These both indicate that the board and PC are communicating.

The Arduino Environment

You now have the Arduino integrated development environment (IDE) installed on your computer and set up to communicate with the Arduino board. With it, you can write code with which to program the microcontroller. In Arduino parlance, these programs are known as sketches.

In essence, the Arduino IDE is similar to a word processor. It comprises four main sections: the menu bar, the toolbar, the text editor and the status area.

The IDE is written in Java and based on Processing, avr-gcc, and other open source software.

Menu bar   Toolbar   Title bar   Tab selector

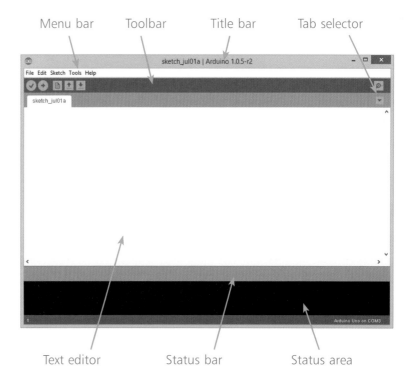

Text editor   Status bar   Status area

Sketches are so called because they are based on the Processing programming language, which lets users create programs quickly in a similar way to scribbling an idea in a sketchbook.

Title Bar
Right at the top is the title bar, which shows the name of the current sketch and the IDE version (Arduino 1.0.5-r2).

Menu Bar
Below the title bar is the menu bar. This offers the following menus:

File Menu
This has a range of commands from the ubiquitous New, Open, Close, etc., to commands specific to Arduino. The latter include:

38

- **Sketchbook** – this shows a list of all the sketches you have created and provides an easy way of opening them.

New	Ctrl+N
Open...	Ctrl+O
Sketchbook	▶
Examples	▶
Close	Ctrl+W
Save	Ctrl+S
Save As...	Ctrl+Shift+S
Upload	Ctrl+U
Upload Using Programmer	Ctrl+Shift+U
Page Setup	Ctrl+Shift+P
Print	Ctrl+P
Preferences	Ctrl+Comma
Quit	Ctrl+Q

- **Examples** – here, you can access a number of pre-written sketches. As they are open source, they are free to use or even to alter for your own purposes.

- **Upload** – clicking this will upload the code in the text editor to the Arduino board.

- **Upload Using Programmer** – this enables you to upload sketches with an external in-system programmer (ISP). We recommend you ignore this option for the time being.

Edit Menu
Again, many of the commands in this menu will be familiar. Those that won't be are:

- **Copy for Forum** – this option copies the code of your sketch to the PC's clipboard in a format that is compatible with the Arduino forum.

Undo	Ctrl+Z
Redo	Ctrl+Y
Cut	Ctrl+X
Copy	Ctrl+C
Copy for Forum	Ctrl+Shift+C
Copy as HTML	Ctrl+Alt+C
Paste	Ctrl+V
Select All	Ctrl+A
Comment/Uncomment	Ctrl+Slash
Increase Indent	Ctrl+Close Bracket
Decrease Indent	Ctrl+Open Bracket
Find...	Ctrl+F
Find Next	Ctrl+G
Find Previous	Ctrl+Shift+G
Use Selection For Find	Ctrl+E

- **Copy as HTML** – enables the code of a sketch to be copied to the clipboard in HTML format, which is suitable for placing in a web page.

- **Comment/Uncomment** – blocks of text that are commented are not uploaded to the Arduino board. Typically, commenting is used to remember or explain how a sketch works.

Sketch Menu
This menu consists of four options that help you to manage your sketches.

There are several reasons why you might want to use an external programmer. These include quicker upload time, lack of a serial connection, or to increase the memory available to your sketch.

…cont'd

Arduino automatically puts every sketch you write in a folder. The "Show Sketch Folder" option opens the folder that contains the current sketch.

- **Verify/Compile** – this option checks the code you have written for errors and then "compiles" it into a format the microcontroller will understand.

- **Add File** – this adds a file to the sketch, which appears in a new tab in the sketch window.

- **Import Library** – enables a library to be added to the sketch. We'll see more on libraries later.

Tools Menu

The tools menu provides various utilities that you may find useful when using Arduino.

- **Auto Format** – auto formats a sketch so it is easier to read.

- **Fix Encoding & Reload** – this fixes typographical errors that can adversely affect a sketch.

- **Board** – enables you to select your board from a list of Arduino boards.

- **Serial Monitor** – displays serial data and is useful for debugging – see Chapter 10.

Toolbar

The toolbar contains icons for the most commonly used commands so you don't have to go looking for them. These are: Verify, Upload, New, Open and Save.

Status Bar and Status Area

The status bar shows status messages regarding current operations. It also acts as a progress indicator so you can see how an upload is proceeding. The status area is used to show error messages.

Text Editor

This is where you enter the code when writing sketches. The editor works in much the same way as a word processor. Right-clicking on text opens an editing menu that offers typical editing options, such as Cut, Copy, Paste, etc.

The Increase Indent and Decrease Indent options are basically formatting commands that enable you to organize your sketches so they can be read more easily.

4 Shields & Libraries

Sometimes you may not want the bother, or may simply not have the time, to design and construct a circuit board yourself. The solution is to use pre-built circuit boards known as shields. For the same reasons, you can also use pre-written sections of code that are known as Libraries.

What is a Shield?

A shield is an extension circuit board that increases the capabilities of the basic Arduino board by simply plugging into its pin headers, i.e. "piggybacking".

Most shields are complete and ready-to-go but some are supplied in kit form, and so need assembling first. It should be pointed out here that many otherwise complete shields are supplied without pin headers – this means that other shields cannot be stacked on top. The solution is to use stacking headers. These are sold as kits that usually include 10-, 8-, and 6-pin headers.

By stacking shields on top of one another, it is possible to incorporate several of them in a project, as shown in the image below. This enables complex projects to be constructed.

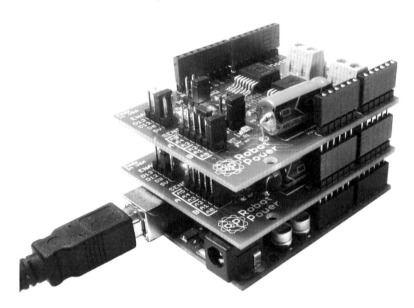

The big advantage of shields for the Arduino beginner is that they enable ready-made circuits to be incorporated into projects without the bother of having to design and build them.

The vast majority of Arduino shields are inexpensive, usually costing between 10 and 30 dollars. Some of the more complex ones cost nearer 100 dollars.

A large number of shields are available covering a wide range of applications. In the next few pages, we'll take a look at some of the most popular ones.

Display Shields

Electronic readouts are an everyday part of our lives – they are used in a host of applications. Both light emitting diodes (LEDs) and liquid crystal displays (LCDs) are used for this purpose. Also common are TFT touchscreen displays. These technologies are available in Arduino shields as we see below:

OpenSegment Shield

This shield offers a four-digit, seven-segment display (each segment being comprised of one LED). It also has an ATmega328 microcontroller that allows you to control every segment individually.

The OpenSegment Shield is available in various colors, including red, blue, yellow and green.

The OpenSegment Shield can be controlled in three ways: Serial TTL communication, SPI serial communication, or I2C serial.

Its features include a selectable baud rate, selectable brightness level, and individual segment control for each digit. It displays numbers, many letters and a few special characters.

TFT Touch Shield

Supplied by Adafruit, this shield offers a large color touchscreen display with a built-in microSD card connection.

The screen measures 2.8 inches diagonally, its backlight is provided by four white LEDs, it has a resolution of 240 x 320 pixels, and offers many different shades for a colorful display.

As a bonus, the display is also resistive, which enables it to detect finger presses anywhere on the screen.

The TFT touchscreen also has a library that detects x, y and z (pressure).

The shield is supplied fully assembled so no soldering is required. Just plug it in and load the supplied library. This can draw pixels, lines, rectangles, circles and text.

Audio Shields

Adafruit Wave Shield

The Wave Shield is an inexpensive board that enables audio to be played with the Arduino. It can play up to 22KHz, 12-bit uncompressed audio files of any length, and provides an onboard DAC, filter and op-amp for good quality audio output.

The shield comes with an Arduino library for easy use; simply drag uncompressed wave files onto the SD card and plug it in.

Audio is played asynchronously so the Arduino can perform tasks while the audio is playing. Volume can be controlled with the onboard thumbwheel potentiometer.

Note that the Wave Shield is supplied as a kit and assembly will require soldering.

MP3 Player Shield

The MP3 Player Shield from Sparkfun effectively turns your Arduino into a fully functional stand-alone MP3 player. As a bonus, it can also decode Ogg Vorbis, AAC, WMA, and MIDI audio files.

Also included is a microSD card reader that simplifies uploading files, and a 3.5mm mini jack to which can be connected stereo headphones, or active (powered) speakers.

The board is supplied fully assembled with the exception of the header pins, which will need to be soldered in place. 6- and 8-pin stackable headers are recommended.

An SD card and loudspeaker are not supplied with the Wave Shield; they must be purchased separately.

Beware

44

Prototyping Shields

Proto Shield Rev 3

A Proto Shield is somewhat different to the usual Arduino shields in that it doesn't actually offer a specific function. Instead, it provides an Arduino-friendly platform that enables circuits of any type to be built. These can be on a permanent basis (by soldering the parts in place), or for experimentation purposes.

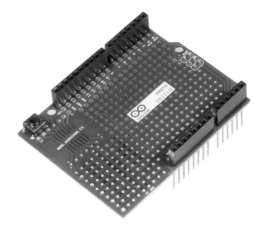

The Proto Shield not only provides extra connections for all the Arduino's input and output pins, it also has enough room to accommodate through hole mounted and surface mounted components.

If bought as a kit, you not only get the board itself, you also get the following parts:

- 1 straight single line pinhead connector 40 x 1
- 1 straight single line pinhead connector 3 x 2
- 2 PCB push buttons
- 1 red LED, 1 yellow LED and 1 green LED
- 15 resistors of various sizes

The Proto Shield kit is supplied fully assembled or as a kit.

ProtoScrew Shield

Very similar to the Proto Shield is the PhotoScrew Shield. The difference is that the PhotoScrew Shield comes with large screw terminals that are connected to the pins. This makes the shield ideal for those who don't like soldering, or need a quick method of assembly/dissembly.

The shield is sold as a kit and will need assembling.

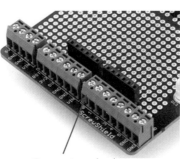

Screw terminals

Hot tip

The Proto Shield enables you to combine custom-built circuits and Arduino into a single module.

Hot tip

Features provided by the Proto Shield include a reset button, a large work area, and an ICSP connector.

Gaming Shields

The Arduino can be used for that most popular of activities, electronic games.

Gameduino

The Gameduino Shield is a games adapter for Arduino. It has a VGA connector that allows video to be sent to any VGA-compatible monitor, projector or other type of display. Sound can be sent via a 3.5mm mini jack.

Other features of the shield include 400 x 300 pixel video output, 15-bit internal color processing, and a 512 x 512 pixel character background. The shield is supplied fully assembled.

With the Gameduino you can either play games or create them. While they are restricted to 8-bits, this is still more powerful than the 8-bit games consoles that were around in the 1980s, such as the Oric-1, Acorn Electron, ZX Spectrum, C64, etc.

A more recent version of this shield is the Gameduino 2. This provides a number of improvements including an integrated touchscreen, making it an all-in-one game controller.

Joystick Shield

While the Gameduino Shield provides the power to play games, it doesn't offer a means of controlling them. Enter the Joystick shield from SparkFun.

This handy board provides all the functions found in a modern games controller. You get an ergonomic control stick with an integrated push button, plus four more buttons that can be used for navigation or game control.

The Joystick Shield is supplied in kit form and so will require assembly.

Hot tip

Amongst many other improvements, the Gameduino 2 offers a 4.3 inch touchscreen, an embedded GPU, an accelerometer and a microSD slot.

GPS Shields

There are many applications these days that require accurate location and/or tracking capabilities. The Global Positioning System (GPS) provides the capability, and the Arduino enthusiast is catered for in this respect by a number of GPS shields:

GPS Shield Retail Kit
This product is supplied in kit form and comes with a GPS receiver as part of the package (another version of this shield is sold without the receiver). You can use different receivers if you want but you will also have to buy and fit the appropriate connectors.

A bonus of the GPS Shield is that it also provides extremely accurate time.

The shield enables you to locate position very accurately and, with it, you can build a range of GPS projects, from tracking your kids' movements, to the creation of location-based services.

The DLINE/UART switch switches the GPS receiver's input/output between Arduino's standard TX/RX pins or any digital pins on the Arduino. Note that the switch must be set to DLINE in order to upload code through the Arduino IDE.

Ultimate GPS Logger Shield
This shield can not only be used for GPS purposes but also to log or save data. This is done courtesy of the integrated SD card slot, which allows you to save data to an SD card.

Other features of the shield include a large surface area for prototyping purposes, and support for an external antenna. The latter is a real boon for projects that will be placed in an enclosure.

The SD card option allows far more data to be stored than would be possible in the Arduino's limited memory.

Simply connect an external GPS antenna to the shield and the module will automatically switch over to use it.

Power Shields

LiPower Shield

If you want to have a truly mobile Arduino, it will need to be battery powered. While you can always use AA or AAA batteries, the LiPower Shield provides a different option – rechargeable lithium batteries.

The problem with these is that they only provide 3.7V. However, the LiPower has this limitation covered with circuitry that boosts the 3.7V to 5V.

Even better, the shield is able to monitor the battery and send an alert when the power remaining reaches a certain level.

The board's built-in charging circuit will charge the battery at 100ma. There is a mini-USB port on the shield that enables you to charge the battery from a USB power source. Alternatively, you can supply a separate regulated 5V source on the "charge" header.

Power Driver Shield

The Arduino does not provide a lot of power so cannot be used by itself for projects that demand a lot of current. The Power Driver Shield provides the solution.

At one end of the board is an ATX power connector. This is the same connector used by computer power supply units. So all you need to do is get hold of one of these units and hook it up to the shield.

At the other end of the shield are six pulse width modulation (PWM) output terminals, which you can use to supply your project with up to 30 amps of current.

Beware

Due to the height of the six power transistors, you will need to double up on the stacking headers if you want to stack a board on top of the Power Driver shield.

Hot tip

The Power Driver Shield is supplied in kit form and will need to be assembled.

Motor Shields

Motors play an important role in many Arduino projects, robotics in particular. Motor shields provide the means of controlling them. We look at two of the most popular below:

Motor/Stepper/Servo Shield

This shield provides virtually everything you are likely to need in a motor shield. It doesn't let you control just one motor but in fact several – up to two 5V servo motors, two stepper motors, or four bidirectional DC motors.

Hot tip

This shield provides thermal shutdown protection.

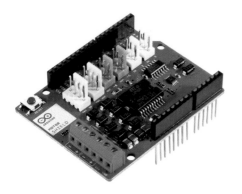

Connecting up the motors is a snap thanks to the large terminal block connectors (these also allow you to power the motors remotely).

The shield can handle 1.2 amps per channel and has a max peak current capability of 3 amps. It features a dedicated driver chip that handles all the motor and speed controls. Only two pins are required to drive the motors.

Motor Shield R3

The Motor Shield R3 is less capable than the Motor/Stepper/Servo Shield mentioned above as it can only control either two DC motors or one stepper motor. However, it can handle currents of up to 2 amps per channel giving a max peak current capability of 4 amps, so more powerful motors can be accommodated.

The board has two channels, each of which use four pins to choose the rotation direction, vary the speed, and sense the current that is flowing through the motor.

Hot tip

The Motor Shield R3 is TinkerKit compatible, which means you can quickly create projects by plugging TinkerKit modules to the board.

The Motor Shield R3 must be powered by an external power supply – a power adapter that provides a voltage between 7v and 12V is recommended.

Communications Shields

Communication is the name of the game these days and Arduino is up to the mark with a number of related shields. Two of the most popular are:

Wi-Fi Shield

The Wi-Fi Shield enables you to connect your Arduino to the Internet wirelessly using the 802.11 wireless specification. This gives Internet related projects the element of mobility.

The shield connects to encrypted networks that use either WPA2 Personal or WEP encryption. It can also connect to open networks.

Features include an onboard microSD card slot, which can be used to store files for serving over the network, or to save search results.

Note that networks must broadcast their SSID for the shield to be able to connect to them.

Ethernet Shield

The Ethernet Shield enables your Arduino to communicate with the Internet directly – a computer is not required. As with the Wi-Fi shield above, it features an onboard microSD card slot that provides data storage facilities.

The shield supports up to four simultaneous socket connections, which connect to the Internet via a standard RJ45 ethernet jack.

The shield also includes a reset controller, which ensures that the board is properly reset when it is powered up. The reset button also simultaneously resets the Arduino board.

SSID stands for "service set identifier". This is basically the IP address of a wireless network.

50

The Ethernet Shield connects to an Arduino board using long wire-wrap headers which extend through the shield. This keeps the pin layout intact and allows another shield to be stacked on top.

Miscellaneous Shields

On pages 42-50, we have reviewed boards from the most popular shield categories. However, there are many others that provide equally useful, if less commonly required, functions. These include:

Weather Shield

The Weather Shield is an easy-to-use board that has built-in sensors providing access to barometric pressure, relative humidity, luminosity and temperature.

It also offers connections that allow you to install optional sensors such as wind speed, direction, rain gauge, and GPS.

The Weather Shield can operate from 3.3V to 16V and has built- in voltage regulators and signal translators. Note that headers, connectors and additional sensors will need to be purchased separately.

Hot tip

The Weather Shield has a temperature accuracy of +/- 3C, a pressure accuracy of +/- 50Pa and a humidity accuracy of +/- 2%.

EasyVR Shield 2.0

Voice recognition technology has improved enormously in recent years – add this capability to your Arduino projects with this voice recognition shield from SparkFun.

EasyVR 2.0 is a multi-purpose speech recognition board designed to add versatile, robust, and cost-effective voice recognition capabilities to virtually any application.

Amongst other features, it offers additional connectors for the microphone input, an 8 ohm speaker output, and an audio headphone jack. A programmable LED is also included to show feedback during recognition tasks.

Hot tip

The EasyVR Shield has 28 built-in speaker-independent commands. These are available in English, Japanese, Italian, French, German, and Spanish.

...cont'd

Arduinos have a limited amount of memory. If large amounts are required, you will need to use a shield such as the MicroSD Shield.

MicroSD Shield

The MicroSD Shield is intended for use in any project that involves large amounts of data, such as audio, video, graphics, data logging, etc. The Arduino Uno (and other Arduino boards) offers a very limited amount of storage space that is unlikely to be enough for these types of application.

The MicroSD Shield provides the solution by equipping your Arduino with mass-storage capability.

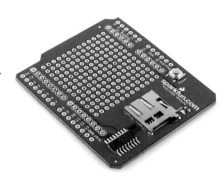

This is the same type of storage as used by digital cameras and MP3 players, and it can add gigabytes of storage space to your projects.

The shield comes with a microSD socket, red power indicator LED, and a reset button. However, it does not come with the headers pre-installed.

Relay Shield

The Relay Shield is a simple board that gives your Arduino the ability to control a single electromechanical relay for switching loads of up to 24V DC or 40V AC, at currents of up to 5 amps.

When operated as a normally-open relay, the Relay Shield is capable of handling loads of up to 10 amps.

The relay is an "SPDT" type, meaning that it can be used either as a normally-open or normally-closed relay.

This means you can choose to either switch on or switch off the load when the output signal from your Arduino is high.

The relay is powered directly from the 5 volt power supplied by your Arduino board, and is compatible with the vast majority of Arduino boards (not just the Uno) that operate at 5 volts.

NFC Shield

Near Field Communication (NFC) is a short-range, low-power wireless link that can transfer tiny amounts of data between two suitably equipped devices held closely together.

The technology is increasingly being used with cell phones and enables them to pay for items by simply tapping the phone on an NFC equipped terminal.

The NFC Shield provides your Arduino with NFC functionality. Not only can it read NFC tags but it can write to them as well.

You can also purchase NFC tags to use with your NFC-equipped Arduino. These come in a range of shapes and sizes.

It can also act as an NFC tag for bi-directional communication with NFC-equipped cell phones and tablets.

The shield is supplied in kit form and so will require soldering.

USB Host Shield

The USB Host Shield allows you to connect a USB device to your Arduino board. It contains all of the digital logic and analog circuitry necessary to implement a full-speed USB 2.0 peripheral/host controller with your Arduino.

With it, you can connect HID devices (mice, keyboards, etc.), game controllers, digital cameras, storage devices (memory sticks, hard drives, etc.), and Bluetooth dongles.

The USB Host Shield is TinkerKit compatible.

Go-Between Shield

This shield is for those situations where you are just not sure if one shield will work with another shield. It is quite common when building Arduino projects to find that problems are caused by two shields both needing the same header pin.

The Go-Between shield can resolve incompatibility issues of this type by quite literally "going between" two shields that aren't compatible and effectively reassigning the pin/pins as necessary.

Before You Buy a Shield

Theoretically, there is no limit to the number of shields that can be stacked on one another. In practice though, there are some limitations you need to be aware of:

- **Dimensions** – you may find that some shields contain components that are taller than the header sockets of the board being plugged into, thus preventing a connection from being made. This problem can be overcome by using stacking headers as spacers.

- **Pin Contention** – make sure the shield does not need a data pin that is already being used by a different shield. This can be difficult to establish and may require you to look at the manufacturer's schematic. Note that it is OK to share the GND, 5V, 3V, RESET, and IOREF pins.

- **Obstruction** – some shields will block access to the inputs or outputs of the shield below them, thus rendering the whole setup useless.

- **Power** – some shields use a lot of power. If you use too many of these, you may find the Arduino board cannot provide enough of it. Shields that are known for their high power requirements are Liquid Crystal Display (LCD) shields and wireless shields. You may also experience power issues with shields that use the 3V supply.

  A simple and easy solution to this problem is to use an external power supply.

- **Interference** – shields that use any type of radio frequency communication can interfere with each other, particularly if they use the same frequencies. Placing shields of this type as far apart as possible may be enough to resolve any issues.

  Another potential problem is electrical interference. This can be caused by electrically noisy boards such as motor driver shields.

- **Software** – shields vary in their software requirements. Combining shields that use large amounts of code in a stack can cause problems by requiring more memory than the Arduino provides, or by initiating resource conflicts.

Hot tip

Always isolate shields that use radio frequencies by placing them as far apart in a stack as possible.

Libraries

An Arduino library is basically a sketch that has been modified slightly so that it can be shared by other people, and also easily updated.

They enable functions to be quickly added to a sketch thus increasing its capability. For example, you could program your Arduino to use a specific type of hardware. Instead of having to write the necessary code yourself, you can simply import it to your sketch in the form of an existing sketch.

You may already have the library in the Arduino software; alternatively you can download it from the Internet. You will find that many other Arduino users have documented the code for a huge number of popular projects and functions. These are freely available for you to integrate into your own sketches.

When you become proficient with programming, you can write libraries yourself. For now, though, we'll take a look at some of the libraries included with the Arduino Uno.

You'll find these by going to Sketch on the toolbar and selecting Import Library. These libraries cover the most popular Arduino subjects and you are sure to find something here that will be useful in one of your own projects.

- **EEPROM** – EEPROM is short for Electrically Erasable Programmable Read-Only Memory. It is a type of non-volatile memory used in computers and other electronic devices to store small amounts of data that must be saved when power is removed. The library enables you to write to and read from the EEPROM component on the Arduino board.

- **Ethernet** – you will use this library when connecting an Ethernet shield to your Arduino. This enables the shield to communicate with the Internet either as a server or as a client.

- **LiquidCrystal** – the LiquidCrystal library enables an Arduino board to control Liquid Crystal Displays (LCDs). The library is based on the Hitachi HD44780 (or a compatible chipset), which is found on most text-based LCDs.

There are a number of reasons for building libraries. These include: simplifying the usage and/or organization of the code; making the code easier to read; and logic decentralization.

Most of the Arduino libraries on the Internet are open-source. This means you are free to use them in your own projects.

...cont'd

- **GSM** – Global System for Mobile Communications (GSM) is an international cellular service available in Europe and other parts of the world. The GSM library enables an Arduino board to do most of the operations you can do with a GSM phone; place and receive voice calls, send and receive SMS, and connect to the Internet over a GPRS network.

- **SD** – the SD library is used with shields that contain an SD memory card. These are used extensively in portable devices, such as cell phones, digital cameras, GPS navigation devices, etc. It enables both reading from, and writing to, SD cards and it supports both FAT16 and FAT32 file systems.

- **Wi-Fi** – when used with the Arduino Wi-Fi Shield, this library allows an Arduino board to connect to the Internet. It can act as either a server accepting incoming connections or a client making outgoing ones. The library supports WEP and WPA2 Personal encryption, but not WPA2 Enterprise.

- **Stepper** – the Stepper library enables you to control unipolar or bipolar stepper motors with an Arduino board. To use it, you will need a stepper motor and the appropriate hardware with which to control it.

- **Servo** – this library is used in conjunction with hobby servo motors. These are motors that have integrated gears and a shaft that can be controlled precisely. The Servo library supports up to 12 motors on most Arduino boards and 48 on the Arduino Mega.

- **Firmata** – Firmata is a standard communication protocol that provides a means of controlling your Arduino from a computer software program. The Firmata library can also be used to selectively send and receive data between the Arduino device and the software running on the host computer.

- **SPI** – the Serial Peripheral Interface (SPI) is an interface bus commonly used to send data between microcontrollers and small peripherals such as shift registers, sensors, and SD cards. With an SPI connection, there is always one master device that controls the peripheral devices. The SPI library allows you to communicate with SPI devices, with the Arduino as the master device.

Hot tip

Wired Equivalent Privacy (WEP) & Wi-Fi Protected Access II (WPA2) are security protocols that secure networks by encrypting transmitted data.

5 Tools & Techniques

The Arduino board by itself is not enough. You will also need tools and equipment, plus knowledge of techniques such as soldering and design.

Circuit Boards

On pages 23-24, we looked at solderless breadboards and how they can be used for quickly building and testing prototype circuits. The drawback with them is that components are easily dislodged as they are not soldered permanently in place.

Therefore, once a breadboard has served its purpose by establishing that a particular circuit is functional, it's then time to transfer the components to something more durable. Here, you have several options.

Printed Circuit Boards (PCBs)

The first is to use a printed circuit board, otherwise known as a PCB. These are constructed of a non-conductive substrate, usually fiberglass, with a copper foil bonded to one or both sides. An etching process removes the unwanted copper leaving conductive copper tracks that form the circuits, as shown in the image below:

Hot tip

Various methods are employed in creating the copper conductive tracks on PCBs. These include silk screen printing, photoengraving, PCB milling and electroplating.

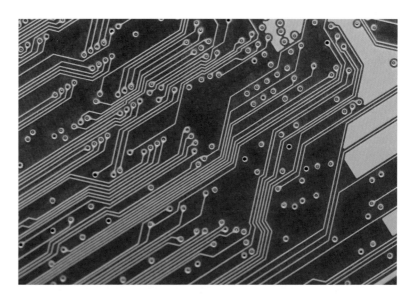

Holes drilled through the board allow components to be connected to the circuits and then soldered in place thus forming a permanent connection.

PCBs can be single-sided (one copper layer), double-sided (two copper layers) or multi-layer. In the latter, the walls of the through holes are plated thus connecting the layers. More advanced PCBs actually have components embedded in the substrate.

Commercially produced printed circuit boards are found in virtually all electrical and electronic equipment these days.

As you would expect, it is inordinately expensive to get one of the commercial firms to make a single board made for you. There is, however, another way. Look on the Internet and you will find a number of online PCB services that will make professional quality boards at an affordable price. One that we know of is at **www. expresspcb.com**

However, it may be that the prices charged by these online services is still more than you want to pay, or that you need the PCB more quickly than they can deliver. If so, you can try the DIY option.

This requires equipment such as a laser printer, transparency film, copper etchant, positive photo resist developer, a blank presensitized PCB board, use of CAD software and a few other bits and pieces.

It's a relatively straightforward procedure that is easily mastered and it will enable you to create custom PCBs of your own design very cheaply indeed. We don't have room here to explain how it is done but there are numerous websites and YouTube videos that you can refer to.

Hot tip

A good description of how to make your own PCB board can be found at www.instructables. com/id/Making-A-Customized-Circuit-Board-Made-Easy

59

Protoboard

A protoboard is a breadboard without the metal clips that hold the components in place. Instead, you solder the component leads on the underside to create a permanent connection.

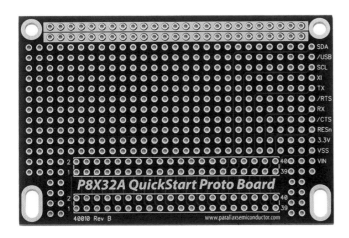

...cont'd

Protoboard is available in all shapes and sizes, and in single, dual and multi-layer forms. You can also get protoboard in the exact size and layout as the Arduino breadboard.

Good quality protoboard comes with the holes through-plated for strength. The plating will also be gold to eliminate the problem of oxidation that can occur with copper plating.

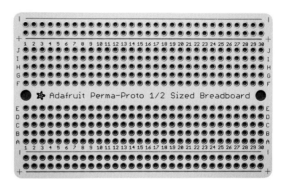

An Arduino type protoboard

As you can see from the image above, it has the same five-hole pad and two power bus lines on each side, design. Also, the markings on the top are identical.

As a result, copying the layout of a working circuit from a solderless breadboard is as straightforward as it can possibly be. Protoboards are also supplied with mounting holes so they can be securely fixed in place.

Stripboard

Also known as veroboard, stripboard is a board that has strips of copper running in one direction all the way down the board as we see in the image below.

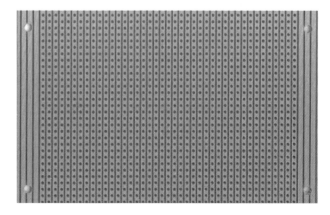

...cont'd

It has pre-drilled holes at regular intervals that allow component leads to be placed in the board; these are then soldered in place. Fixing holes are located in each corner.

Once the components are in place, a knife or scalpel is used to cut the copper tracks where necessary in order to form the circuit.

Perfboard

In perfboard, holes are pre-drilled at standard intervals across a grid. These are usually spaced 0.1 inches apart. Each hole is surrounded by a round or square copper pad.

In this type of board, the components are soldered in place but the connections between the components can be made in more than one way.

One method is the wire-wrap technique, in which special sockets that have long tail pins are used to hold the components in place. These tail pins protrude from the other side of the board and circuits are formed by connecting the pins with wire. To make a connection, several turns of wire are taken around the pin hence the term "wrapping".

A special motorized tool is usually employed for wrapping although it can be done with a hand tool. Not only does it make an extremely reliable connection, the wrapping process also holds the component sockets firmly in place.

However, this technique is not used much these days. Most people who use perfboard solder the components in place and connect them with wire cut to size.

Hot tip

Wire-wrap joints are the most reliable type of joint. Soldering, however, is quicker and easier, while still providing a good joint.

Soldering

As you can see from the preceding pages, it probably won't be long before you need to learn the technique of soldering, as just about the only type of circuit board that doesn't require it is the solderless breadboard. The push-fit type of connection used in these may be quick and convenient but it does not provide a reliable fixing. A technique that does is soldering.

What is Soldering?

Soldering is a way of joining metal. It works by melting a metal with a low melting point (solder) on to the metals being joined. When the solder cools and hardens, the metals are thus joined. In a circuit, two metals are joined – the lead of the component and the copper track on the board.

A soldered joint is quick and simple to make, and will permanently bond a component to a circuit board with a strong and electrically conductive fixing. Another advantage is that it is always possible to undo a soldered joint by reheating it should, for example, a component need to be replaced for some reason. For these reasons, soldering is the most common method of securing electronic components to circuit boards.

You will need to acquire a number of items before you can get started. The first of these is the soldering iron.

Soldering Irons

When looking for a soldering iron, you have four types from which to choose:

- Fixed temperature
- Variable temperature
- Portable
- Solder stations

Fixed temperature irons – the most simple soldering iron is the fixed temperature type. You get an iron with a power cord, a sponge to clean the tip, and a bit of metal on which to rest the hot iron while not in use.

Don't forget

Soldered joints can be "unmade" as easily as they are made.

An important specification to take note of is the iron's power rating (wattage). You will see models rated from 12W right up to 50W. The higher the wattage, the hotter the iron. So what do you need in this respect?

When soldering a joint, it is important that the solder melts before the heat from the iron spreads and causes damage. Too low a temperature is just as likely to cause this as is too high a temperature.

So the answer is a midrange iron in the region of 25W or so. This will be ideal for the type of components you will be using in your Arduino projects.

We recommend buying a soldering iron with a power rating of around 25 watts.

Variable temperature irons – variable temperature soldering irons are basically the same as fixed temperature models apart from the fact that their power is adjustable. Typically, these offer a power rating between 0W and 50W.

The power rating given to the iron will be the highest wattage it is capable of. If you buy this type of iron, it makes sense to get one with a high rating as this will provide a better range of temperatures.

The advantage of variable temperature irons is that they can be used for a wide range of soldering applications – small low temperature components such as diodes and transistors, and high temperature components such as large resistors and transformers.

Portable irons – portable irons do not have a power cord; instead they are powered either by battery or by gas canister, both of which are situated inside the handle.

Battery models use standard AAA batteries and the gas models use butane gas.

...cont'd

The advantages of these irons are that they can be used in locations where no electrical power is available, and in tight spaces where a cord might be a hindrance.

Disadvantages are that they are expensive, the gas models are potentially dangerous, and with both the battery and gas models the power source doesn't last too long.

Solder stations – At the top end of the soldering iron range are the soldering stations. These consist of a soldering iron, a docking base for the iron complete with an integrated sponge compartment for cleaning the iron's tip, and a variable power supply unit.

Hot tip

It can be very useful to have a portable soldering iron as well as a corded model.

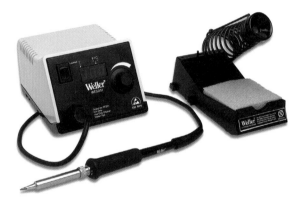

Better quality stations offer features such as microprocessor controlled temperature, digital temperature readout, and auto power-off.

The main advantage of solder stations is the same as for variable temperature soldering irons – the ability to provide a range of operating temperatures. Beyond that, the digital readout models allow extremely precise temperature control, plus the stations provide other features as mentioned above.

The disadvantage of course is the cost – solder stations, particularly at the top end, are expensive.

Solder

The next item on your shopping list is solder which, as we have already seen, is a fusible metal alloy used to secure components to a circuit board.

Hot tip

Steer clear of Solder Stations until you know for sure you will need the features they provide.

There are many different types of solder but for use on circuit boards, there are two main types: lead and lead-free.

It comes in a range of thicknesses starting at around .02 inches, which is ideal for circuit board use, to much thicker stuff that is used to connect heavy duty components, e.g. speakers.

Solder with an approximate thickness of .02 inches is perfect for soldering components in circuit boards.

Most solder is made from a combination of tin and lead – the proportions are roughly 60% tin and 40% lead. This is known as lead solder and one of its advantages is that it has a low melting point, which means you can use a cooler iron and thus be less likely to damage your components. The disadvantage, of course, is that lead is highly poisonous so you may want to give lead solder a miss for this reason.

The alternative is lead-free solder. This is comprised of any of a number of metals, such as tin, copper, silver, bismuth, indium, zinc and antimony. However, while it may be safer to use, lead-free solder has a higher melting point than lead and, also, it does not create as reliable a joint.

Lead solder is potentially harmful and, as lead-free solder works nearly as well, we recommend you use the latter.

The final thing you need to know about solder is that in order to create a good joint, you need a solder flux. This helps to overcome the effects of oxidation that can prevent the solder from flowing freely.

Most solder these days is sold with a hollow core that contains the flux. As you apply the solder you will see this burning off. However, very cheap solder may be sold without the flux core and, in this case, you will need to buy a tin of flux and apply it by hand.

Flux also helps to distribute the heat from the soldering iron.

...cont'd

Soldering Iron Tips

The business end of any soldering iron is its tip and, over time, these items degrade. For this reason they are replaceable. Another reason is that soldering iron tips come in an assortment of shapes and sizes, and so enable a single iron to be used for more than one type of job.

These range from pointed conical tips that are used for soldering small closely spaced components to flat screwdriver shaped tips that are suited to soldering large components.

Soldering Iron Accessories

A very useful soldering iron accessory is a holder or stand that provides a safe place to keep the hot iron. Many irons will come with one and we recommend that the one you buy does as well.

Also useful is a small sponge or scouring pad with which to wipe excess solder from the tip periodically.

For those of you who envisage doing a lot of soldering, or find it difficult to hold more than one item at the same time, a desktop vise (also known as a third hand) will be a definite boon.

These hold the circuit board for you by means of alligator clips, allowing you to get on with the soldering. Some models, like the one on the right, also provide a magnifying glass for close-up work.

Desktop vises can be screwed or bolted to the work surface so the circuit board can be securely held at any angle.

When you make the inevitable mistake and solder the wrong component to the board, or in the wrong place, you will need to undo the joint by reheating it and removing the solder. Doing this without over-heating the component is difficult. To this end, you will find a solder sucker to be very handy.

This device is basically a small vacuum pump that sucks molten solder away from a joint leaving it clean so the component can be removed.

The devices can be a bit fiddly to use as you have to operate them with one hand while keeping the solder molten with the other.

Another tool that is used for removing solder is the solder bulb. This is simply a hollow rubber bulb with a nozzle that sucks up the solder when the bulb is released.

An alternative to the solder sucker is solder wick. This is copper wire that has been braided into strips and is supplied in reels. It works by providing a large surface area for molten solder to adhere to.

Just hold the solder wick on the joint being disassembled and press down on it with the soldering iron until the solder melts and is absorbed by the wick. Then pull the wick away and the solder will come away with it. You may need to repeat the exercise to remove all the solder.

Other Tools

Apart from the items already mentioned, you will need a pair of wire strippers as shown below:

This tool is invaluable for removing the insulation from insulated wire quickly and precisely.

...cont'd

Also handy will be a pair of needle-nose pliers or tweezers. These enable you to hold small parts and to manoeuvre them in confined spaces.

Finally, consider getting a pair of wire snips. This will be useful in removing the excess lead wire from a soldered joint.

Safety

Soldering creates fumes that potentially pose serious health risks if not managed properly. Lead-based solder gives off lead oxide fumes that can cause lead poisoning. Also poisonous are the fumes given off by rosin-based flux.

To protect against the dangers posed by solder fumes, it is important to work in a well ventilated area. Alternatively, use an exhaust fan to vent the fumes either outside or with a filter. Many soldering stations provide an integrated fan and this is a very good reason to pay the extra dollars required to get one.

However, while there is no dispute regarding the health risks posed by soldering, it must be said that there is absolutely no danger associated with occasional soldering work. It is only an issue for those who do a lot of it.

Probably more likely, potentially, is the risk of damaging your eyes. Soldering irons can "spit" tiny pieces of solder in all directions and while these are too tiny to cause any serious damage, again, prolonged exposure won't do your eyesight any good. So while it may seem excessive, we recommend you use goggles when soldering.

On a slightly different tack, you may also consider the safety of your components by using an anti-static strap. This will stop them being damaged by the electrostatic electricity in your body.

Soldering Technique

When you actually get down to the mechanics of soldering, you will find it's not as easy as it looks. Probably the first problem you will encounter is getting the solder to melt so that it flows freely.

The cause of this is almost certain to be oxidation on the soldering iron's tip, which reduces its ability to transfer heat. You can always spot a tip that is oxidized – instead of being shiny silver as it should be, it will be a dull dark color. It is very difficult to solder when a tip is in this state.

1. So before you start, check the tip. If it isn't clean, wipe it with a sponge a few times

2. Next, you need to "tin" it. Tinning is a procedure that will prevent the tip from being re-oxidized for a while. Do it by melting solder on to the tip until it is coated with the solder – it will now be bright silver. Then wipe off the excess. A tinned tip should be good for a number of joints before it needs to be retinned

The image shows a badly oxidized tip on the left and on the right, the same tip after it has been tinned.

3. You are ready to begin soldering. Before you do though, remember this simple rule – solder always flows towards heat. So heat up the component lead and the circuit, i.e. the joint, before applying the solder. If you don't, the solder just won't stick. Do not load the iron with solder and then try to transfer it to the joint – it won't work

4. Insert the component's leads through the holes in the circuit board, flip the board over and bend the leads back so the component stays in place

Hot tip

If you have trouble getting the solder to melt, try using a thinner solder.

Hot tip

Some soldering irons are sold with tips that have a special anti-oxidation coating.

Hot tip

New iron tips should always be tinned before use.

...cont'd

5 Now, place the tip of the hot iron at the point where one of the leads touches the circuit board pad. Hold it there for a couple of seconds to heat the lead and pad, and then touch the solder to them; do not touch it to the tip of the iron. The solder will melt and flow into the joint. When you see a nice neat blob of shiny solder on the joint, lift the iron away – the joint is made

Beware

Be careful not to overheat a joint – this may damage the component. Soldering a joint should take no more than a couple of seconds. If necessary, let the joint cool down and then try again.

6 Give it a second or so to cool and then take a close look at it. The blob of solder should cover the entire pad without running over on to another pad. If necessary, remove the solder with a solder sucker and start again

7 Finally, use your snips to cut away any excess lead wire

With a good joint, the blob should be shiny silver. If it has a dull appearance you probably have what's known as a cold joint. This can produce a poor electrical connection between the component and the board, which can give rise to a variety of problems.

The principle cause of cold joints is not applying enough heat to the joint so the solder doesn't flow freely. If a joint does look poor, simply reheat it until the solder remelts and, if necessary, add a little more solder.

Power Adapters

In many situations, you will be able to power your Arduino board from your computer via the USB cord. However, should you ever have to situate the board away from the computer, you will need to use a power adapter.

The question of precisely *which* power adapter to use is a common one on the Arduino forums. The answer is that any adapter that meets the following criteria will be suitable:

- The adapter must provide a direct current (DC) output.

- The adapter must provide a 2.1mm center-positive plug.

- The adapter's current output must be at least 250mA.

- The output voltage should be between 6 volts and 20 volts. However, be aware that if it is less than 7V it is possible that the board's 5V pin may supply less than 5V, resulting in instability issues. If it is more than 12V, the voltage regulator may overheat and damage the board. The recommended range is therefore 7 to 12 volts.

Usually though, any "wall wart" with a center polarity positive connector and an output of 6V to 20V will work fine. It doesn't matter if the adapter's output is unregulated as the Arduino comes with an onboard voltage regulator.

Note that it is also possible to supply power to the Arduino board via the voltage in pins (Vin).

The Arduino board will take a wide range of voltages. However, the recommended voltage is between 7 and 12 volts.

Test & Diagnostic Equipment

Sometimes it simply doesn't work at all; sometimes it works but not as it should, or not very well; sometimes you just aren't sure *if* it will work.

In the latter scenario, you need to check things out before going any further and, in the others, you need to establish the cause of the problem.

To help you do so, a number of test and diagnostic devices are available. These range from simple circuit testers to advanced equipment such as oscilloscopes and signal generators.

Circuit Tester

This is a simple and inexpensive device that is used to check for the presence of electricity in a circuit board. It consists of two leads connected to a neon lamp.

To use it, touch a live wire with one lead and ground with the other lead. If the neon lamp lights, it confirms that power is present. If the light doesn't come on, either the power is off or you have a faulty circuit.

Continuity Tester

Another simple device that is nevertheless surprisingly useful, the continuity tester consists of an indicator (either a light or a buzzer) and a source of electrical power, usually a battery. Each end of the device acts as a probe.

In operation, you must first switch off the power to the circuit. Then connect the clip to one end of the circuit to be tested and touch the probe to the other end. If the circuit is good, electricity from the battery flows through it and activates the indicator.

Hot tip

Circuit testers can also be used to check a piece of equipment is grounded.

Hot tip

Continuity testers are useful for checking switches.

72

Multimeter

A multimeter is an electronic measuring device that provides a number of measurement functions in one unit. These include the ability to measure voltage, current, and resistance.

Multimeters can be analog (these use a microammeter whose pointer moves over a calibrated scale) or digital (these display the measured value in numerals). Of the two, digital meters are by far the most common these days.

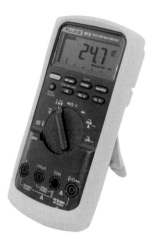

The devices are available as small hand-held units that can be used for fault-finding in electrical circuits, or much more advanced bench models, which can measure to a high degree of accuracy, plus provide a greater range of measurement functions.

A decent multimeter is the only piece of test equipment the average Arduino user is likely to need.

For the electronic hobbyist, the hand-held units are ideal as they provide pretty much everything you are likely to need at a very inexpensive price. Bench models provide features you are not likely to need and can cost thousands of dollars.

Oscilloscope

An oscilloscope is an advanced type of test and diagnostic device that can display the waveform of an electrical signal at any point in a circuit. They can even display non-electrical signals, such as sound or vibration, by converting them to a voltage.

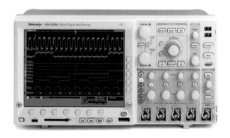

The waveform can be analyzed for such properties as amplitude, frequency, rise time, time interval, distortion, etc.

Oscilloscopes are large bulky devices that are only suitable for bench use. They are also very expensive and require the user to have an advanced knowledge of electronics in order to operate them. For the hobbyist, they are not really an option.

Design Software

Years ago, before the advent of digital technology, planning a circuit board layout was done with pen and paper. There is still nothing wrong with this approach; however, these days you can simplify the process by using a computer program.

Some of these programs also act as a simulator and show you exactly what the circuit does. In other words, you can see if what you've designed will actually do what you intend it to.

Having established that the circuit works, you can then save it as a file for yourself, to share with others or to send to a PCB manufacturer.

There are a number of circuit design programs, many of them free. A popular one is Designspark, available at **www.rs-online.com/ designspark/electronics/**

An example design is shown below:

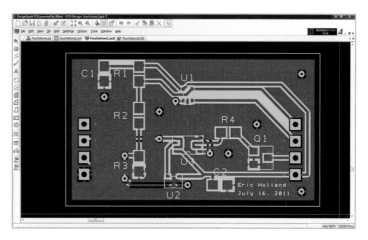

Other well known programs include:

- Fritzing (**www.fritzing.org/home**)
- Kicad (**www.kicad-pcb.org**)
- Expresspcb (**www.expresspcb.com**)
- Pad2pad (**www.pad2pad.com**)
- FreePCB (**www.freepcb.com**)
- CadSoft EAGLE (**www.cadsoftusa.com**)
- 123D Circuits (**www.123dapp.com/circuits**)

All these programs are free to use although, with some, the free versions come with limitations.

123D Circuits is an interesting option as it allows you to design your circuits online – there is no need to install a program on your computer. Furthermore, it offers an Arduino option – this enables you to not only design Arduino circuits but also to program the microcontroller with code and then test that it all works by running a simulation. All this can be done without touching the physical Arduino board at all.

123D Circuits is one of our favorite circuit design programs and is highly recommended.

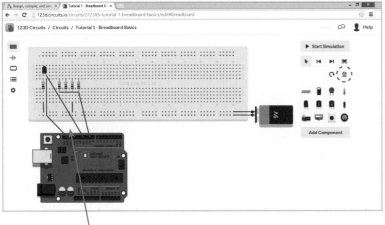

Design your circuit by dragging components from the pallet

Program the microcontroller by entering code in the editor

A viewing option in 123D Circuits lets you see the design as it would look as a PCB board.

Any errors in the code will be highlighted just as they would be in the code editor on your computer. When everything is working as it should, simply copy the design to the physical board.

Schematic Diagrams

When you sit down to design a circuit, draw a schematic diagram. This will be invaluable both in the planning and construction stages.

A circuit consisting of just a handful of components really does not need to be designed as such – being so simple, you will have a mental picture of the required layout. However, this will not be the case with more complex circuits that require many parts – you will quickly lose track of things.

This is where schematic diagrams come into play. Basically, they act as a road map by displaying all the components in the circuit and showing clearly how they are connected.

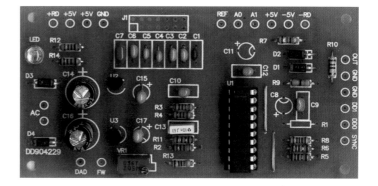

Take the circuit shown above. Even though it is relatively simple, it would be very difficult to design (or build) without some sort of reference – it would be all too easy to make connection errors.

With a schematic, however, you can see exactly what goes where. This simplifies both the designing of the circuit and its construction.

A variation of the schematic diagram is the block diagram. This is used when large circuits are planned with each block representing a section of the circuit.

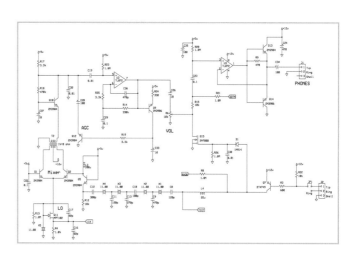

In a schematic, each type of component is represented by a symbol, so to be able to understand what you are looking at (and also how to create a schematic yourself) you need to know what these symbols are.

Also, commonly used components are usually given an identifier. This helps to differentiate between them on circuits that contain many of them.

Component Symbols

Many different types of components are used in electronic circuits and each type can be subdivided into more types. For the time being, we are just going to look at the symbols for the most commonly used components.

The Resistor

This part is found in virtually all circuits and does what the name suggests – it resists the flow of electricity. The most common type is the fixed resistor. Each resistor in a circuit will be given an identifying label, i.e. R1, R2, R3, etc.

Resistor

Resistors are rated in Ohms.

The Capacitor

Another common component, the capacitor is often used in conjunction with a resistor to form a timing circuit. It can also be used as a filter, e.g. to block DC signals but pass AC signals. The device is labeled with the letter C, i.e. C1, C2, C3, etc.

Capacitor

Capacitors are rated in Farads.

The Diode

Diodes allow electricity to flow in only one direction – from the anode to the cathode. The anode is on the left (the triangle) and the cathode (the vertical line) on the right, so the current flows from left to right. The device is labeled D1, D2, D3, etc.

Diode

Diodes (and other semiconductors, such as transistors) are rated in several ways.

Hot tip

Some components also have their value denoted on circuit boards.

As with diodes, LEDs and transistors are semiconductor devices.

...cont'd

The Light Emitting Diode (LED)

LEDs work in much the same way as other members of the diode family. However, they also convert electricity to light. The symbol is the same as for an ordinary diode but for the two diagonal arrows pointing away from the anode. As with diodes, LEDs are labeled as D1, D2, D3, etc.

LED

The Transistor

Transistors amplify current. They can also be used as switches. The vertical line at the top of the symbol represents the collector, the vertical line at the bottom represents the emitter, and the horizontal line at the left represents the base. The downward pointing diagonal arrow indicates a NPN transistor. If it was pointing the other way, it would be a PNP transistor. The devices are labeled as Q1, Q2, Q3, etc.

NPN transistor

Connections

In a circuit, wires often cross – sometimes they touch and make a connection, sometimes they don't. In a schematic, this is indicated as follows:

Crossing without touching – when two wires cross without touching each other, there are two ways of representing it. Neither is incorrect – it's just a matter of preference.

Non connected crossed wires

Crossing and touching – when two wires cross and touch thus making a connection, the schematic shows it as a dot at the point of connection, as shown below:

Two connected wires

Ground

All electronic circuits need a return path for electricity to flow back to the other side. This function is provided by a circuit's Ground pin, and every circuit will have at least one of these.

Ground

Grounding is also a safety measure in high-voltage circuits.

6 Electronic Components

By itself, the Arduino board cannot do much. In order to create useful projects with it, you will need to expand it by building external electronic circuits. In this chapter, we look at the electronic components supplied with the Arduino, the various types of each and explain what they do.

Supplied with Arduino

Included in the kit supplied with your Arduino board are a number of electronic parts as shown below:

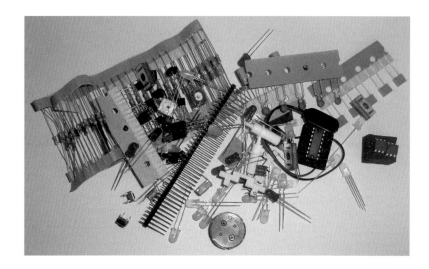

Hot tip

The components supplied with your Arduino are just to get you started. In time, you will need to buy more.

The full list is comprised of the following:

- 6 photoresistors
- 3 potentiometers
- 10 push buttons
- 1 temperature sensor
- 1 tilt sensor
- 1 alphanumeric LCD
- 1 white LED
- 1 RGB LED
- 8 red LEDs
- 8 green LEDs
- 8 yellow LEDs
- 3 blue LEDs
- 1 6/9V DC motor
- 1 servo motor
- 1 piezo capsule
- 1 H-bridge motor driver

- 2 optocouplers
- 5 transistors
- 2 mosfet transistors
- 5 100nF capacitors
- 3 100uF capacitors
- 5 100pF capacitors
- 5 diodes
- 3 transparent gels
- 1 male pins strip
- 20 220 ohm resistors
- 5 560 ohm resistors
- 5 1 Kohm resistors
- 5 4.7 Kohm resistors
- 10 10 Kohm resistors
- 5 1 Mohm resistors
- 5 10 Mohm resistors

These get you off to a good start by enabling you to immediately start building some simple circuits. In this chapter, we will take a look at some of these parts, the various types of each part, and see what they do.

Resistors

Of all the many different types of electronic component, resistors are the most common and will be found on literally every circuit board, no matter what its use.

The purpose of these devices is simple – it's to resist the flow of current in a circuit, and the ability to do this is measured in ohms. Resistors are available in many values, each providing a different level of resistance.

There are three main types of resistor – fixed resistors, variable resistors and specialized resistors.

Fixed Resistors
This is the most common type and it has three main uses:

- **Protection** – fixed resistors are used to limit the current flowing through other components, thus preventing them from damage. Protecting LEDs is a typical example of this.

- **Splitting** – fixed resistors are used to split the voltage between different sections of a circuit.

- **Timing/delay** – when used in conjunction with a capacitor, a fixed resistor forms a timing or delay circuit.

Variable Resistors
These are resistors in which the level of resistance can be altered.

With the standard variable resistor, the resistance is continuously variable within a set range. Examples of their use is a radio's volume control, and bass and treble controls in Hi-Fi equipment.

A different type is the potentiometer. These have a control that is usually adjusted just once to set a value in a circuit and then left at that level. This type of resistor is not accessible to the user.

Specialized Resistors
More specialized types of resistors include the thermistor where the resistance is sensitive to changes in temperature. These are used in digital thermometers, 3D printers, toasters, hair dryers, etc.

Another is the photoresistor; these are similar to thermistors except they are sensitive to light rather than temperature. An example of their use is controlling shutter speed in cameras.

The basic function of a resistor is to restrict the flow of current in a circuit. This property can be used in a number of ways.

In electronic circuits, the properties of two or more different components can be combined to produce a specific action. A resistor and capacitor in tandem will form a timing circuit.

Resistor Color Coding

Many resistors are too small to have the resistance value printed on their bodies. Therefore, bands of color are used to represent the resistance value.

The first and second bands represent the numerical value of the resistor, and the third band specifies the multiplier. The color bands are always read from left to right starting with the side that has a band closest to the end.

The color chart below shows the various colors used and what they signify:

Beware

Most resistors are tiny in size. You will probably need a magnifying glass to read the color bands.

Hot tip

Resistor color code converters can be found online. Simply enter the various colors to find the resistance.

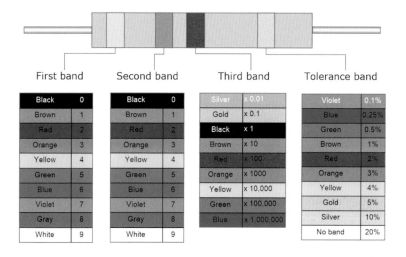

First band   Second band   Third band   Tolerance band

First band		Second band		Third band		Tolerance band	
Black	0	Black	0	Silver	x 0.01	Violet	0.1%
Brown	1	Brown	1	Gold	x 0.1	Blue	0.25%
Red	2	Red	2	Black	x 1	Green	0.5%
Orange	3	Orange	3	Brown	x 10	Brown	1%
Yellow	4	Yellow	4	Red	x 100	Red	2%
Green	5	Green	5	Orange	x 1000	Orange	3%
Blue	6	Blue	6	Yellow	x 10,000	Yellow	4%
Violet	7	Violet	7	Green	x 100,000	Gold	5%
Gray	8	Gray	8	Blue	x 1,000,000	Silver	10%
White	9	White	9			No band	20%

First band – the band on the left is the first band and it indicates units of 10. Each unit is assigned a specific color.

Second band – the second band also indicates units of 10. These are assigned a specific color as well. So the yellow and violet bands on the resistor above indicate a value of 47.

Third band – the third band shows the level of multiplication. In our example above, the multiplier is red (x 100) so the resistor has a value of 47 x 100 – 4,700 ohms or 4.7 Kohms.

Fourth band – the fourth band indicates the tolerance. This is a percentage of the quoted resistance. A brown band indicates a tolerance of 1%, a red band 2%, an orange 3% and so on. If there is no band at all, this indicates a tolerance of 20%.

Hot tip

It is impossible for manufacturers to give a resistor an exact value. So they provide a tolerance factor. For example, a 10 ohm resistor with a tolerance of 10% will have an exact value of between 9 and 11 ohms.

Capacitors

Capacitors are the next most common component in electronic circuits. They come in a range of types, shapes and sizes.

Essentially, these devices are comprised of two conductive plates that are separated by an insulator.

When an electric current is applied, a charge builds up between the two plates. By then turning off the supply of current, this energy can be stored in the device for a specific period or until an external event triggers its release.

The unit of measurement of capacitance is the Farad. This is actually a very large unit so most capacitors are rated in micro-farads (uF), nano-farads (nF) and pico-farads (pF).

There are many different types of capacitor, the main ones being film, electrolytic, ceramic and variable.

Film Capacitors – film capacitors are one of the most common types and can be considered a general purpose capacitor as they are inexpensive and suitable for a wide range of applications.

Electrolytic Capacitors – electrolytic capacitors are used when high levels of capacitance are required in a circuit. These devices are polarized, i.e. they have a positive and a negative pole. This means they must be connected the right way round.

Ceramic Capacitors – another common type, these capacitors have a high level of capacitance relative to their size.

Variable Capacitors – also known as trimming capacitors, these provide an adjustable level of capacitance. They are used to tune circuits or for compensation purposes.

Capacitors have many uses. These include filtering (removing unwanted signals), smoothing (evening out fluctuations in a signal), DC blocking (capacitors only pass AC current), voltage snubbing (limiting the effects of voltage spikes), pulsed power (capacitors are effectively a tiny battery), tuning circuits (variable capacitors are used to tune radios, for example), and sensing (capacitors can be used to sense changes in the environment).

Hot tip

If the capacitor body is large enough, its full rating will be printed on it. If not, you will see two or three numbers.

If it's just two, it indicates pico-farads (pF). For example, 50 would be 50pF.

If there are three, the third number is a multiplier. For example, 503 would indicate 50 x 3 giving a rating of 150pF.

Hot tip

Electrolytic capacitors are large devices that resemble a barrel.

Beware

Working with electrolytic (polarized) capacitors can be dangerous. If they are connected the wrong way round, they can self-destruct in an alarming fashion, i.e. explode.

Inductors

An inductor is basically a coil of wire in the shape of a cylinder or a toroid (donut), which is wound around a central core. When a voltage is applied to the device, a magnetic field develops around the coil. An electric current is stored within this field. Another property of the inductor is that it resists any change in the amount of current flowing through it.

Inductance is thus defined as the size of the electromotive force (emf), or voltage, induced in a component (especially coils) compared with the rate of change of the electric current that produces the emf or voltage.

The unit of inductance is the Henry. This is a large unit and more common ones are the micro-henry (uH), the milli-henry (mH) and the nano-henry (nH).

Inductors tend to be classified by the material used for the core. Typical examples are:

Air Core Inductors – these inductors don't have a core at all, the lack of which produces a low level of inductance. They are commonly used in high frequency applications, e.g. TV, radio.

Iron Core Inductors – these produce a high level of inductance and so are used in applications that require a lot of power.

Ferrite Core Inductors – ferrite is found in many different types of inductor and these can be used for a variety of purposes.

Iron Powder Inductors – iron oxide allows a high level of inductance to be produced in a small area.

Typical uses of inductors include:

Radio frequency (RF) chokes that squelch high frequency signals and thus prevent them from interfering with other circuits.

Tuned circuit applications such as filters and oscillators. In this role, inductors are combined with capacitors to produce a resonant frequency.

Protecting sensitive equipment from destructive voltage spikes and high currents.

Hot tip

The shape of the core is significant. Toroidal cores provide more inductance for a given core material and number of turns than solenoidal (rod-shaped) cores.

Hot tip

Inductors are sometimes classified by their mechanical construction. For example, bobbin inductors, toroidal inductors, ceramic inductors and film inductors.

Hot tip

Factors that affect inductance include the number of turns in a coil, the length of a coil, the cross section area of a coil, and the material used for the core.

Diodes

A diode is a polarized component that allows current to move through it in only one direction. It has two leads: one forming the anode and the other the cathode. The cathode is usually indicated by a silver band as in the example below:

When the voltage at the anode is higher than it is at the cathode, current will flow from the anode. However, if the voltage is higher at the cathode, the diode will not allow current to pass.

These devices are semiconductors and, as such, do not have a specific rating or value as do resistors (resistance), for example. Instead, semiconductors have various parameters specified, such as peak inverse voltage (PIV) in the case of diodes.

Diodes come in many types, the most common of which are:

Light Emitting Diodes (LEDs) – constructed of a material that produces light in a process known as electroluminescence, the ubiquitous LED is found virtually everywhere these days.

Zener Diodes – unlike standard diodes that will not let current flow in the reverse direction, zener diodes will but only at a specified level. This property makes them ideal voltage regulators.

Photodiodes – these devices sense light and are able to convert it into a current. They have a huge number of applications, particularly in consumer electronic devices.

Rectifiers – a rectifier is an array of diodes arranged in such a way that they convert alternating current (AC) to direct current (DC). A bridge rectifier that employs four diodes is a common type.

Diodes can be used in many different ways. One of the most common is as a source of reliable and inexpensive light (LEDs).

Another is their ability to convert AC to DC, which is used in power adapters for electronic devices such as tablets.

The light sensing properties of photodiodes are used in many different ways. These include light meters, x-ray detectors, light pens, infrared remote controls, smoke and flame detectors, etc.

Diodes only pass current in one direction.

Diodes are constructed from semiconductor material. This is a substance that has a conductivity between that of an insulator and of metal, either due to the addition of an impurity or because of temperature effects.

It forms the basis of modern electronics as it is used in integrated circuits (ICs), diodes and transistors.

Transistors

Transistors are devices constructed from a semiconductor material and that have a minimum of three terminals. The basis of operation is that a voltage applied to one pair of terminals alters the current at another pair of terminals.

Bipolar transistor   Field effect transistor

The three terminals of a bipolar transistor are called the emitter, the collector and the base. Current flows between the emitter and the collector while the base is used to control the device.

This transistor is the key component in virtually all modern electronics. Not only does it provide two important functions (switching and amplification), it is also very cheap to manufacture. Furthermore, they can be manufactured in incredibly small sizes. For example, the processors used in computers can contain several billion transistors.

Being made from semiconductor material, transistors are easily damaged if connected incorrectly. To help in identifying them, transistor leads usually have a dot near the collector and/or a tab near the emitter.

There are many different types of transistor but they can all be put in one of two main groups:

Bipolar Transistors – bipolar transistors are current controlled devices that have three regions: emitter, collector and the base. They are often used for amplification purposes.

Field Effect Transistors – field effect transistors are voltage controlled and also have three regions: source, drain and the gate. These devices are often used as digital switches.

In analog circuits, transistors are used as amplifiers, while in digital circuits they are used as switches. These are their two main uses; however, they can also be used as circuit buffers and regulators.

86

Transistors can also be classified by function. For example: switching transistors, power transistors, high frequency transistors and phototransistors.

Relays

Relays are electrically operated switches that consist of several parts including an electromagnet, an armature and a set of contacts. In more specialized relays, the electromagnet is replaced with a solid-state device. Some relays will have just one set of contacts while others will have several (the latter are used where a number of circuits need to be controlled by one signal).

In operation, the electromagnet is energized by the application of a current, and produces a magnetic field. The magnetic field then attracts an armature, which in turn moves a contact thus completing a circuit.

A useful property of relays is that the input circuit is completely separate from the output circuit. For this reason they can be used to electrically isolate circuits for protection purposes.

The main use of relays is to take a low current and use it to switch a much larger current on or off. Their ability to do this is incredibly useful with devices such as sensors, which produce levels of current that are too low to power a circuit. The relay acts as an interface and protects the sensor from the high current in the load circuit.

In situations where really large currents need to be switched, relays are often cascaded, i.e. a small relay switches the current needed to drive a larger relay that then switches the current needed to power the circuit.

Common types of relay include:

Electromechanical Relays – consisting of an electromagnet, an armature and sets of contacts, these have two important limitations – slow speed and a relatively short lifetime.

Reed Relays – while these are also electromechanical devices, due to their design, they are some ten times faster than the electromechanical relay. They also have a longer lifetime.

Solid-State Relays (SSRs) – constructed from semiconductor material, these devices have no moving parts. As a result they are much faster and longer lasting than electromechanical relays.

Typical uses of relays include automobiles where 12V DC is used to control high current circuits, such as headlights and starter motors, and motorized home appliances such as food mixers.

Reed relays are prone to damage from arcing. This is when an arc jumps across the contacts and melts a section of a contact's surface.

Transformers

Transformers are devices that can change the voltage of an alternating current (AC). A step-up transformer increases the voltage while a step-down transformer decreases it.

These devices consist of two coils of insulated wire wound around a core. An input voltage is applied to one coil (called the primary), which produces a constantly varying magnetic field around the coil. This magnetic field, in turn, induces an alternating current in the output coil (called the secondary). From there it is passed to a circuit.

The voltage at the secondary coil is dependent on the number of turns in the coils. For example, if there is one turn in the primary coil and 10 turns in the secondary coil, the voltage in the secondary will be 10 times that in the primary. If there are 10 turns in the primary and one turn in the secondary, the voltage in the secondary will be one tenth of that in the primary.

Some transformers have more than one secondary coil enabling them to supply a number of load circuits.

There are a number of transformer types, all of which operate on the principles outlined above. These include:

Power Transformers – these are widely used to convert wall power to the low voltages required to power electronic devices.

Pulse Transformers – these produce electrical pulses that have uses in digital and telecommunication applications.

RF Transformers – RF transformers have many uses in high frequency applications such as VHF, UHF and microwave.

Audio Transformers – this type of transformer is used in audio circuits, e.g. filtering unwanted interference to the audio signal.

The examples above are just some of the ways in which transformers are used. There are many others.

Motors

Electric motors are ubiquitous – they're in your home, at work and just about everywhere else you can think of. These devices convert electrical energy to rotational energy.

They consist of a number of parts that include a casing or housing, rotor, stator, commutator, shaft or axle, and brushes.

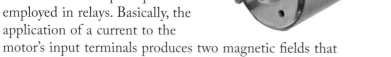

Motors use the same electromechanical principles employed in relays. Basically, the application of a current to the motor's input terminals produces two magnetic fields that alternately push and pull the shaft. This causes the shaft to rotate.

The rotational force or torque created by a motor is determined by three things: the magnetic strength of the stator (a permanent magnet that surrounds the rotor), the strength of the electric current at the input terminals, and the number of turns in the rotor (the more turns, the faster it rotates).

Electric motors can be grouped in three main types: alternating current (AC) motors, direct current (DC) motors, and universal motors. AC motors need AC to run, DC motors need DC to run and universal motors can run on both AC and DC.

Universal motors are most commonly found in powerful household appliances that are typically used for short periods. For example: food processors, blenders, and vacuum cleaners.

AC motors are available in single-phase and three-phase versions. Single-phase AC is what is typically supplied in a home, while three-phase AC is used in industrial applications.

DC motors are available in a number of types. These include brush motors, brushless motors, and stepper motors. Of these, brush motors are by far the most common.

A stepper motor is a type of brushless motor that allows a computerized control system to "step" the rotation of the motor. This is useful in applications where something needs to be moved a specific distance. Robotics is a typical example.

Electric motors are rated in horsepower, e.g. 1/2 horsepower.

The speed of an electric motor is usually specified as rotations per minute (RPM) at no load condition.

Hot tip

Two common uses of ICs are the digital watch and the calculator.

Hot tip

Integrated circuits can also be configured as a surface-mount chip with 8, 14 or 16 pins.

Integrated Circuits (ICs)

Integrated circuits are the basis of modern microelectronics. These devices are constructed from a semiconductor material such as silicon, and are commonly referred to as chips or microchips.

They can contain literally billions of components such as diodes, transistors, and resistors. As a result, they form highly complex circuits that are nevertheless very small in size. Other advantages include:

- **Weight** – ICs are very light devices

- **Cost** – it costs much less to make an IC than it would to use the standard components required to provide the same function

- **Reliability** – as there are no soldered connections, few interconnections and low temperatures, failure rate is very low

- **Power** – because the components are microscopic in size, ICs use very little power

Most integrated circuits are packaged in a plastic casing with a number of legs or pins protruding for connection to the circuit. These pins are arranged in a dual-in-line (DIL) configuration and are numbered in an anti-clockwise direction.

DIL integrated circuits can be plugged into a IC socket. This eliminates the need for soldering, which can damage them. It also allows ICs to be easily replaced; during a repair for example.

High power integrated circuits can generate a lot of heat so they have a metal tag that can be connected to a heatsink to dissipate the heat.

Because they contain entire circuits, virtually all ICs are designed to provide a specific function. Analog ICs are used in amplifiers, radio receivers, voltage comparators, voltage regulators, etc.

Digital ICs are mostly used in computers. Typical functions include timers, counters, calculator chips, memory chips, microcontrollers etc. The microcontroller on your Arduino board is a digital IC.

Sensors & Actuators

Probably the most basic Arduino project is the blinking LED. This is good as an introduction to the subject but nothing else – it serves no useful function, which is the whole point of Arduino.

To enable your Arduino projects to actually do something, it is usually necessary to incorporate electronic sensors and actuators.

Sensors

Sensors are devices that detect a physical change in their environment, e.g. temperature, motion, light, etc., and either generate an electrical signal (passive sensors) or alter a property such as resistance level in the device (active sensors). These signals are proportional to the degree of change and provide the input to a circuit.

Most active sensors require an external power supply (called an excitation signal) in order to operate. Passive sensors, on the other hand, do not need a power source as they generate an electrical signal in response to an external stimulus.

There are many different types of sensor that can be used to detect a huge range of properties. These include acceleration, light, color, radiation, sound, temperature, humidity, touch, liquids, and distance to name just a few.

Actuators

In a circuit, the signal produced by the sensor is processed and then sent to an external device to perform a specific function. The device is known as an actuator and it converts electrical or magnetic energy into a different type of energy. For example, motion, light, heat, etc.

Any device that is powered by one type of energy and produces a different type at its output can be classified as an actuator. The electric motor is one of the most well known types of actuator.

Sensors Commonly Used With Arduino

There are sensors available that can detect just about every property known to man. This means that the scope of your projects is limited only by your imagination and ingenuity.

To begin with, however, you will probably restrict yourself to more basic projects that, typically, will use the following sensors:

Hot tip

Sensors can be either analog or digital.

...cont'd

Force Sensitive Resistor

Pressure Sensors

There are various types of pressure sensor used in Arduino projects but one of the most common is the Force Sensitive Resistor (FSR). This is used to detect physical pressure such as pushing, squeezing and weight.

In operation, the resistance of an FSR is proportional to the pressure applied to it. The greater the pressure, the lower the resistance.

Photoresistor

Light Sensors

Types of light sensor include photovoltaic cells, photoresistors, photodiodes and proximity light sensors. Photovoltaic cells come in large panels that are not suitable for Arduino projects.

The others, however, are. Due to their instant response, photodiodes are used as switches in digital circuits, e.g. remote controls. Proximity sensors respond to infrared light and are used in robotics, while photoresistors are commonly used for gauging and responding to light levels.

Thermistor

Temperature Sensors

As with other sensors, there are various types of temperature sensor. Two of the most common are the thermistor and analog sensors.

With the thermistor, changes of temperature alter the resistance of the device. A typical application is in digital thermostats. Analog temperature sensors are solid-state devices and are used when precision is required. They are also reliable and inexpensive.

PIR

Motion Sensors

The most commonly used type of motion sensor with regard to Arduino is the Passive Infrared (PIR) sensor. These devices detect changes in infrared radiation. They are small, inexpensive and simple to use. A popular use is in home security systems.

Other types of motion sensor employ microwaves and ultrasonic sound waves. Increasingly popular is the use of video cameras to detect motion in their field of view with suitable software. This enables the recording of video to be triggered by the detection of motion.

7 Circuits

This chapter is a primer on the subject of electrical circuits. You will learn important principles such as Ohms Law, the different types of circuit and current, how to use resistors and capacitors, and much more.

Concepts of Electricity

What is Electricity?

All objects are composed of minute building-blocks known as atoms. Every atom, in turn, consists of three smaller components known as particles. These are the proton, the neutron and the electron.

The more positive protons or negative electrons there are in an atom, the stronger the attraction and thus the stronger the flow of current. This attraction is known as the "charge".

Electrons contain a negative charge, protons a positive charge and neutrons are neutral – they have neither a positive nor a negative charge.

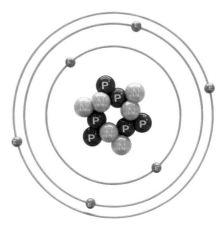

The protons and neutrons are tightly bound together in the nucleus of the atom and so are highly resistant to movement. The electrons are not bound together and spin around the nucleus at a distance. As a result, they are easily dislodged from their orbit, i.e. they are much less resistant to movement.

Atoms contain equal numbers of protons and electrons, thus they are balanced and stable. However, if an atom loses an electron it then has more protons than electrons, and is positively charged. An atom that gains electrons on the other hand, has more negative particles and is thus negatively charged. If any atoms in the vicinity are also charged, either positively or negatively, interaction will take place between them.

This takes the form of a flow of electrons from one atom to another. One electron leaves its atom and joins its neighbor leaving the original atom short of an electron. This in turn attracts an electron from a different atom. This negative to positive flow of electrons is called a current – what we know as electricity.

The flow of electrons (electricity) is always from negative to positive.

Conductors & Insulators

The degree to which electrons can move about in an atom varies according to the material. With some types of materials, such as metals, the electrons in the atoms are so loosely bound to the nucleus that very little force is required to move them.

In other materials, such as glass, the atom's electrons are bound much more tightly to the nucleus and so they need much more force to make them move.

The degree to which electrons can move in a material is known as electric conductivity. This is determined by two main factors: The first is the type of atom in the material and the second is how the atoms are linked.

Materials in which electrons can move easily are known as conductors, while materials in which the electrons are tightly bound (few or no free electrons) are called insulators. Good conductors include metals (silver being the best of all), mercury, graphite, and water. Good insulators include glass, rubber, oil, ceramic, wood and air.

Hot tip

Some materials experience changes in their conductivity when heated or cooled. For example, when heated to a very high temperature, glass becomes a good conductor.

Circuits

Creating a flow of electrons is one thing; employing it in a useful manner is another. The first thing we need to do is find a way of directing the flow to where we want it, i.e. control it.

This is done by creating an unbroken physical path for the electrons so they have no option but to go where they are directed. We do this by building circuits.

Constructed of a conductive material (typically copper) that aids electron flow, circuits form a closed loop in which various components are placed to modify properties of the electric current, such as strength, frequency, polarity, etc.

This is demonstrated in the diagram below.

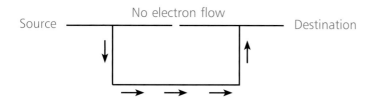

Hot tip

It's important to be aware that electron flow will cease whenever the circuit is broken. It doesn't matter where the break is.

From this we can see that when there is an unbroken path between the electron source and the electron destination, the electron flow is obliged to take this route. If there is a break in the circuit, however, they cannot flow.

...cont'd

Voltage

The need for a continuous path isn't the only requirement for electrons to flow, however. It is also necessary to initiate the electron flow, i.e. get it going, and to keep it flowing once it has started. In other words, there has to be some force, or pressure, to physically force the electrons around the circuit.

We do this with batteries and electrical generators. In the case of batteries, these devices consist of an anode, a cathode and electrolyte. With all batteries, the anode is the source and the cathode the destination.

A chemical reaction in the battery generates a build-up of electrons at the anode. This results in an electrical difference, or imbalance, between the anode and the cathode. In order to restore the balance, the electrons at the anode try to get to the cathode but cannot because the electrolyte stops them. Thus, a force, or energy, has been created.

Because this energy cannot be released, it is known as *potential* energy or, more commonly, voltage. However, if a wire is connected between the anode and the cathode, a circuit is formed thus releasing the energy. Current will flow and will continue to do so as long as the battery is producing voltage.

If the circuit is broken at any point, though, the flow of current will stop immediately – it will not be present in any part of the circuit.

The strength of the electron flow, and hence the strength of the current, is proportional to the voltage applied to a circuit. Connect two 1.5 volt batteries in series and the voltage across the circuit will be 3 volts, and the current will be twice as high as it would be with one 1.5 volt battery.

Resistance

When electrons flow through a conductor, they encounter natural resistance, which obstructs and so reduces the flow. This resistance is present in every type of material with some offering much more than others.

Basically, it is caused by collisions between moving electrons and other elements known as ions.

Hot tip

If a circuit is broken, the voltage at the battery will also be present between the two ends of the break. The end connected to the positive side of the battery (the cathode) will also be positive and the end connected to the negative side (the anode) will be negative. This is known as polarity.

The level of resistance offered by a conductor is determined by several factors:

- **The atomic structure of the conductor** – in some materials, e.g. metals, the electrons can move much more freely than in others, e.g. glass. These materials offer much less resistance to current flow.

- **The length of the conductor** – the resistance of a long wire is greater than the resistance of a short wire because electrons collide with ions more often.

- **The thickness of the conductor** – the resistance of a thin wire is greater than the resistance of a thick wire because the former has fewer electrons to carry the current.

The phenomena of resistance has two unwanted effects:

- **Heat** – when moving electrons collide with other elements in a conductor, the energy expended is released as heat. This can cause damage to components and be dangerous to users. It adds another variable that needs to be taken into account when designing circuits.

- **Energy Loss** – energy lost as heat is energy wasted. As energy is expensive, this loss needs to be minimized. Also, as with the need to deal with heat, energy loss is a factor that can have implications in circuit design.

It's not all bad news though. Resistance of specific values can be deliberately introduced into circuits in the form of *resistors*. These components play a very important role in electronics as they enable the flow of current in a circuit to be precisely controlled.

Resistance is a measure of how much an object opposes the passage of electrons. The unit of electrical resistance is the ohm.

The energy expended by moving electrons in overcoming the resistance in a material is lost as heat.

Voltage Drops

It's not just resistors that oppose the flow of current. In fact, all electronic components have an inherent resistance to it. In order to overcome this resistance, an electric current has to expend energy across the component in question.

This expenditure of energy causes a reduction in voltage that is described by the term "voltage drop".

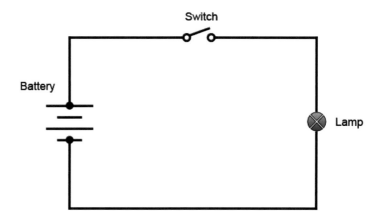

A voltage drop is the amount of voltage that is used as electrons pass through a resistance. All the voltage in a circuit is used, i.e. the sum of the all the voltage drops will equal the source voltage.

98

As an example of this, consider the simple circuit shown above. This comprises a 9 volt battery, a lamp and a switch. With the switch open, the voltage across the battery terminals is 9 volts.

When you close the switch and thus complete the circuit, the lamp lights up. If you were to measure the voltage across the battery now, you'd see that it has dropped to approximately 7.5 volts – a difference of 1.5 volts.

Measuring a voltage drop is done by measuring the voltage before it enters the load and then measuring it as it leaves the load. The difference between the two is the voltage drop.

This drop of 1.5 volts is a voltage drop and it is caused by the energy the battery has to expend in order to overcome the resistance of the lamp. All components in a circuit, even the wires, present a certain amount of resistance and so will cause an associated voltage drop.

Many applications will work just fine with a range of voltages. Others, though, are very supply voltage sensitive. This means all the potential voltage drops have to be taken into account during the design stage and the supply voltage adjusted accordingly.

Power

Resistance, voltage and current aren't the only quantities active in circuits; there is another one, which is called power. Power is basically the measure of the amount of work that can be done in a specific period.

In an electric circuit, it is a function of both the voltage and current in that circuit. However, it is not proportional to the current and voltage but rather is equal to current multiplied by the voltage. Accordingly, power is expressed by this formula, where P = power, I = current, and V = voltage:

$$P = I \times V$$

The unit of measurement for power is the Watt.

It should be clearly understood that power is the combination of the voltage and current in a circuit. Think of it in this way: Voltage is the specific work per unit charge, while current is the rate at which electric charge travels along a conductor.

Therefore, because voltage is analogous to the work done in lifting a weight against gravitational pull and current is analogous to the speed at which that weight is lifted, together voltage and current constitute power.

Hot tip

The unit of electrical power, the watt, can be considered analogous to the unit of mechanical power, Horsepower. 1 horsepower is equal to 746 watts.

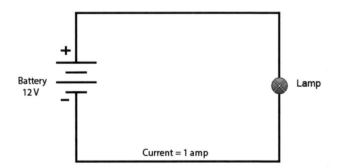

Battery 12 V

Lamp

Current = 1 amp

Don't forget

If there is no voltage or current in a circuit, there can be no power.

In the simple circuit above, we have 12 volts and a current of 1 amp. The formula P = I x V tells us the lamp is releasing 12 watts of power (this will be in the form of heat or light, or both).

If the circuit is broken at any point, current stops flowing and so there will be zero power. Similarly, if the voltage fails for any reason, again, there will be no power.

Series & Parallel Circuits

So far in this chapter, we have shown you some simple circuits. These actually come in two types: series and parallel. We'll now explain the difference between them and show some typical uses for each.

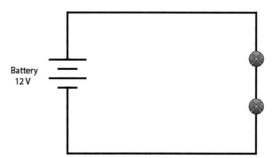

Hot tip

In a series circuit, the current is equally shared by the components. The voltage drops across each component will differ according to each component's resistance but they will total the supply voltage. All the resistances add up to equal the total resistance.

In a series circuit, all the components are connected end-to-end so there is just a single path for electricity to flow through. This is demonstrated in the circuit above, which comprises a battery and two lamps.

The parallel circuit shown below contains exactly the same components but they are wired in a different manner.

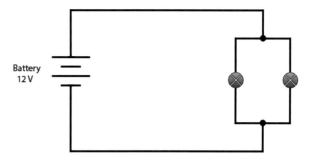

Hot tip

In a parallel circuit, all components share the voltage equally, the currents in each branch add up to the total current, and the resistances drop to equal the total resistance.

Instead of being connected end-to-end so that the current has to flow through one to get to the other, each lamp has its own branch, i.e. is independent of the other. This has two effects:

● In the series circuit, the current is shared between the lamps. So the more lamps there are, the dimmer each will be. In the parallel circuit, however, each lamp gets the full current and, as a result, they glow more brightly than in the series circuit. No matter how many lamps there are, they all glow brightly.

● In the series circuit, if one of the lamps burns out the circuit is broken and so the flow of electricity stops. In the parallel circuit this doesn't happen because regardless of which lamp burns out, there is still an available path through the other lamp. The level of illumination may have halved but there is still some available.

In the real world, however, very few circuits are so simple. Most in fact, are combination circuits that contain elements of both series and parallel as in the example below:

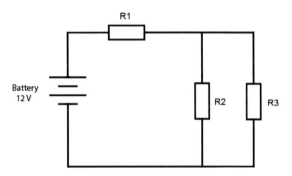

Hot tip

In a combination circuit, the power source and control or protection devices are usually in series; the loads are usually in parallel.

In this circuit, current flows from the battery through R1 then splits with equal amounts going through R2 and R3, and then rejoins before flowing back to the other side of the battery.

Common Uses for Series Circuits
Series circuits tend to be simple with relatively few components. One of the best known examples is the old type of Christmas tree lights where all the lights go out if just one bulb fails.

Another is a flashlight, which has a battery, switch, wires and a bulb, all connected in series.

Common Uses for Parallel Circuits
Parallel circuits are found everywhere. The lighting circuit in your house is parallel as is the lighting in the street outside. The high, medium and low controls in an electric heater put two elements in parallel for full power, one element by itself for half power and two elements in series for low power.

Batteries connected in parallel produce more current. A typical application for this is boat engine ignition systems.

Beware

Every time a component is added to a parallel circuit, more current is drawn from the supply. Thus it is possible to overload a circuit by requiring more current to flow than the circuit can safely handle.

Series Circuits & Ohm's Law

We have seen that the flow of electrons along a conductive path is called a current. We have also seen that the force required to force electrons to flow is called voltage. Furthermore, the flow of current depends not just on the voltage across a circuit but also on the amount of resistance in that circuit.

Thus, there is clearly a fundamental relationship between these three quantities. Specifically, for a given resistance, current is directly proportional to voltage. In other words, if you increase the voltage through a circuit whose resistance is fixed, the current goes up. If you decrease the voltage, the current goes down.

This relationship is known as Ohm's Law and it is expressed in the mathematical formula shown below:

$$V = I \times R$$

From this we can see that voltage is equal to current multiplied by resistance. So if you know the current and resistance in a circuit, you can use these two quantities to calculate the voltage. Furthermore, by changing the formula slightly, we can also calculate the current and resistance.

$$I = \frac{V}{R} \qquad R = \frac{V}{I}$$

Dividing the voltage by the resistance gives the current, while dividing the voltage by the current gives the resistance. Let's see how this works in practice.

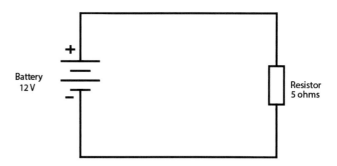

In the circuit above, the current is calculated by:

$$I = \frac{V}{R} = \frac{12 \text{ volts}}{5 \text{ ohms}} = 2.5 \text{ amps}$$

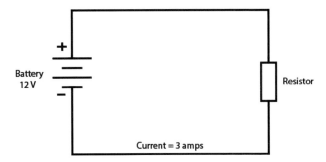

In this circuit, we know the voltage is 12 volts and the current is 3 amps. The value of the resistor is calculated by:

$$R = \frac{V}{I} = \frac{12 \text{ volts}}{3 \text{ amps}} = 4 \text{ ohms}$$

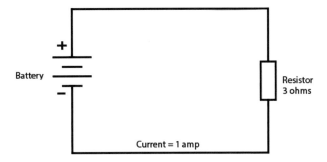

Finally, to calculate the voltage assuming the resistance and current are known:

$$V = I \times R = 1 \text{ amp} \times 3 \text{ ohms} = 3 \text{ volts}$$

Ohm's Law is a very important concept in circuits and will be frequently used. Memorizing the three formulas is highly recommended.

Parallel Circuits & Ohm's Law

The Ohm's Law equations can also be applied to parallel circuits. However, due to the different way of laying out components in these circuits, the equation has to be modified.

The modification is only necessary though, when calculating resistance values – current and voltage are calculated in the same way as they are in series circuits – I = V/R and V = I x R.

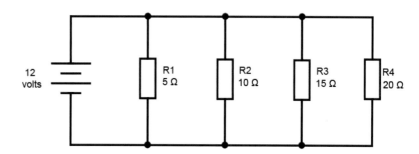

Hot tip

The voltage across resistors connected in parallel is the same for each resistor.

The circuit above has four resistors connected in parallel. If they were connected in series we would just add the values together to get the total. In a parallel circuit this doesn't work. To get the total resistance we need to use the reciprocal formula shown above-right. This works as follows:

$$R_T = \frac{1}{\frac{1}{R1} + \frac{1}{R2} + \frac{1}{R3} + \frac{1}{R4}}$$

$$R_T = \frac{1}{\frac{1}{R1} + \frac{1}{R2} + \frac{1}{R3} + \frac{1}{R4}} = \frac{1}{\frac{1}{5} + \frac{1}{10} + \frac{1}{15} + \frac{1}{20}}$$

$$\frac{1}{0.2 + 0.1 + 0.06 + 0.05} = \frac{1}{0.416} = 2.4 \text{ ohms}$$

Don't forget

When calculating the total resistance in a parallel circuit, you cannot just add up the values of the resistances as you would in a series circuit.

So, four resistors that connected in series would total 50 ohms, present a total resistance of 2.4 ohms when connected in parallel. This can be explained as follows:

Let's assume a circuit with two 4 ohm resistors connected in parallel branches. This circuit will offer two equal paths for the flow of electrons with half the charge passing through the resistor in one branch and the other half through the resistor in the other branch.

...cont'd

Therefore, only half of the total charge in the circuit will encounter the 4 ohm resistance of either branch. Thus, as far as the battery producing the charge is concerned, the two 4 ohm resistors in parallel are the equivalent of one 2 ohm resistor in the circuit.

Using the formula on page 104, any number of resistors in a parallel circuit can effectively be reduced to just one. Once this single value is known, it is then possible to calculate the voltage and current as described on pages 102-103.

Things get much more complicated when dealing with combination circuits that include both series and parallel elements. The trick is to isolate the parallel elements (and there may be several) and work out the total resistance of each with the formula on page 104. Then repeat the exercise until you effectively have just one element. Now all you have to do is add the resistance of this to the series resistances to get the total resistance in the circuit.

For example:

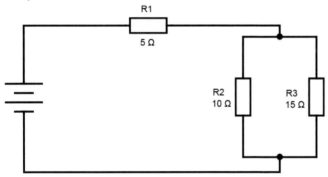

In this circuit R2 and R3 are in parallel. The first step is to work out the total resistance of the two resistors.

$$\frac{1}{\frac{1}{10} + \frac{1}{15}} = \frac{1}{0.1 + 0.06} = \frac{1}{0.16} = 6.2 \text{ ohms}$$

R2 and R3 effectively form a single resistor with a value of 6.2 ohms. As this is in series with R1, simply add the two to give a total circuit resistance of 11.2 ohms.

Hot tip

Once you have calculated the total resistance in a parallel circuit, you can then work out the voltage and current as per a series circuit.

Resistance in Circuits

We have seen what resistors are, the various types of resistor, how they work, and how to work out their value from resistor color codes.

We'll now look at some common ways they are used in circuits:

Protection

All electronic parts are built to specific tolerances. If these tolerances are exceeded, e.g. a part is exposed to excessive current, it may be damaged leading to possible failure of the circuit. Resistors provide the solution.

In the circuit below-left, we have an LED connected directly across a 9 volt battery. The current will be far higher than the LED is rated to withstand and it will burn out immediately.

Hot tip

When connecting a resistor, you don't have to worry about polarity. Current can pass through them in either direction.

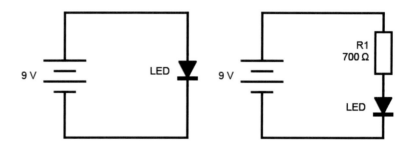

To prevent this, we place a 700 ohm resistor before the LED as shown in the circuit on the right. This will limit the current to a level that the LED can handle – typically 10 to 20 mA.

To work out the required resistance, we use Ohm's Law. Therefore, we need to know both the voltage and current in the circuit.

Don't forget

Remember that Ohm's Law works with fundamental units, i.e. ohms, volts and amps. In our example here, 10 mA has to be converted into 0.01 A.

LEDs have a voltage drop of some 2 volts across them (this information will be in the LED's specifications) so the total voltage is 9 - 2 giving us 7 volts. We need a current of about 10 mA (also in the LED's specifications) so the equation will be:

$$R = \frac{V}{I} = R = \frac{7}{0.01} = \textbf{700 ohms}$$

Voltage Dividers

A voltage divider is a circuit in which two resistors are used to convert a high voltage into a lower one.

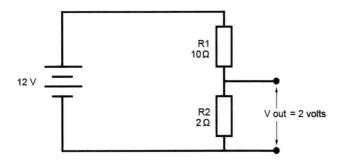

Hot tip

If the two resistors are equal in value then the output voltage is half that of the input. This is true regardless of the resistors' values.

Voltage dividers are simple. In the circuit above, current flows from the battery to R1. At R1, 10 volts is dropped leaving 2 volts to be dropped at R2. As the output is taken from across R2, it will thus be 2 volts.

If you were to change the value of R1 to 5 ohms and the value of R2 to 7 ohms, the output would

$$V_{out} = V_{in} \times \frac{R2}{R1 + R2}$$

now be 7 volts. The output of any voltage divider circuit can be set with the mathematical formula shown above.

Hot tip

The equation states that the output voltage is directly proportional to the input voltage and the ratio of R1 and R2.

Timing/Delay Circuits

When used in conjunction with a capacitor, a resistor forms a timing or delay circuit.

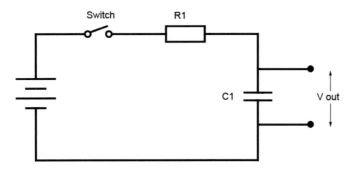

When the switch is closed, current flows through the resistor and begins charging the capacitor. As the capacitor charges, it allows current to flow to the load circuit (V out). When it is fully charged, current to the load circuit stops.

The time between current starting and stopping is a "delay", the period of which can be set by the values of the resistor and the capacitor. It can be used to trigger external events on and off.

Hot tip

In a timing circuit, the value of the resistor determines the rate at which the capacitor charges and thus the length of the delay.

Capacitance in Circuits

As with the resistor, capacitors are used in virtually all electrical circuits. Basically, these devices store electrical charge in much the same way as a battery does, and this property can be employed in many ways. We saw on page 107 how they are used in conjunction with resistors to form a timing circuit; we'll now look at some other uses for this versatile component.

Capacitive reactance is the internal resistance of a capacitor to the flow of current.

Filtering

All capacitors present an inherent internal resistance to the flow of current – this is known as capacitive reactance. Unlike fixed resistors where the resistance doesn't change, a capacitor's reactance varies according to the frequency of the applied signal.

As the frequency increases, the reactance decreases and more current flows through the capacitor. As it decreases, the reactance increases thus restricting current flow. Effectively, the capacitor acts as a variable resistor, the resistance changing as the frequency of the applied signal changes.

Therefore, a capacitor (in conjunction with a resistor) can be used to pass or block specific bands of frequencies, in which capacity it is called a filter. High Pass filters pass high frequencies, Low Pass filters pass low frequencies, and Band Pass filters pass frequencies in a specific range, or band.

A High Pass filter is shown below:

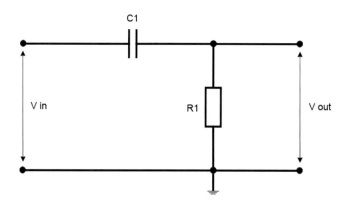

Basically, a filter is a circuit that will reject all unwanted frequencies in an electrical signal. Only the ones required will be passed.

In this type of filter, the current flows through the capacitor first while, with the Low Pass filter, the capacitor and resistor positions are reversed so that current flows through the resistor first.

...cont'd

Smoothing

Another very important use for capacitors is in power supply circuits where alternating current is converted to direct current. Typically, these circuits employ a rectifier diode that basically chops off the negative (lower) half of the AC signal. However, this still leaves the positive (upper) half, which will vary greatly from zero to maximum.

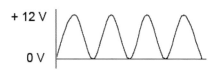

The waveform above shows a rectified AC signal that varies continuously between 0 volts and 12 volts – no good at all for a circuit that requires a current at a constant level. The solution is to place a smoothing capacitor across the rectified output as shown in the circuit below:

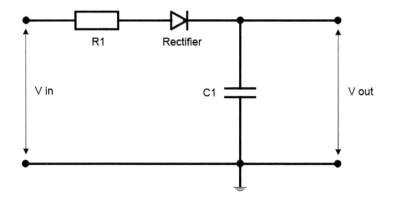

C1 is the smoothing capacitor. When the output from the rectifier rises, the capacitor charges up. When the rectifier output drops, the capacitor discharges thus maintaining a flow of current. The input and output waveforms are as shown below:

Electrolytic capacitors are usually used in smoothing circuits.

The greater the amplitude of the current fluctuations and thus the greater the waveform, the larger the required capacitor.

Alternating & Direct Current

There are two types of current – direct current (DC) and alternating current (AC). Both have advantages and disadvantages that make them suitable for some purposes and less so for others.

Direct Current (DC)

With direct current, the flow of electrons is constant and in one direction. This is why it is called direct current. Plotted on a graph it looks like this:

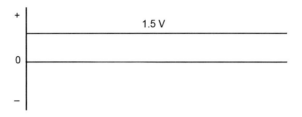

DC is used in most digital circuits as it is ideally suited for low-power, short-range applications. It is found in most electrical appliances, computers, game pads, etc.

Alternating Current (AC)

With alternating current, the flow of electrons is not constant in one direction as with DC. Instead, it changes direction periodically, flowing from positive to negative then back to positive and so on. An AC waveform is shown below:

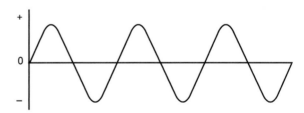

AC comes in a variety of waveforms. The most common is the sine wave shown above. Also common is the square wave, which is used in digital circuits.

AC is more suited to high-power, long-range applications such as industrial power supplies. It is also easier to produce than DC. In many low-power applications, such as computers and home appliances, it is converted to DC by a circuit within the appliance or power adapter.

Hot tip

Direct current can be produced in several ways. These include solar cells, rectifiers and batteries. Alternating current is produced by generators and alternators.

Hot tip

In the United States the AC power supply switches direction 60 times a second, i.e. a frequency of 60 Hertz.

8 Programming Arduino

The ability to program is an essential part of creating projects with your Arduino board. In this chapter, we examine how it's done, taking a look at the most common elements and commands used in writing program code. These include functions, loops, variables, conditional statements, constants and more.

Programming Concepts

Left to itself, the microcontroller on the Arduino board will do absolutely nothing – you need to give it instructions. These instructions are known as sketches, and the act of writing them is known as programming.

There are a number of programming languages, the most popular of which include Java, Python, Ruby, PHP, C, and C++. A simplified version of C and C++ is the language used for programming Arduino microcontrollers.

If you already know C, you are good to go. If you don't, though, worry not – relatively few commands are necessary in order to write sketches. They can be grouped into four categories:

- Operation commands that evaluate expressions, do arithmetic, etc.

- Jump commands that cause a program to jump immediately to another part of the program.

- Branch commands that evaluate a condition and jump if the condition is true.

- Loop commands that repeat a section of code a specified number of times.

This chapter explains what these commands are and how to use them to program your Arduino.

It's important to realize at the outset that the Arduino will do exactly what it is instructed to do – nothing more and nothing less. It is only aware of what is in the sketch – it has absolutely no knowledge of the programmer's intentions!

Furthermore, the sketch's code must be formulated using a specific syntax if the Arduino's microcontroller is to interpret it correctly.

A program built-in to the Arduino IDE, which is known as a compiler, reads the language specific text written in the editor and converts it into machine language, or code.

This is then uploaded to the microcontroller, which runs the sketch by executing the code line by line.

Most programming languages contain a relatively small number of commands. Complex operations are possible due to a computer's ability to combine and repeat instructions millions of times a second.

A well known saying in computer programming is "Garbage in, garbage out."

Comments

Comments are just that – comments made by the person writing the sketch as a reminder of something, or for the benefit of other people. This can be to explain what the sketch, or a specific section of it, does or how it works.

They are ignored during the compilation process and therefore have no effect on the sketch. Comments can be either single-line or multi-line.

Multi-line or block comments are enclosed within the symbols /* and */. They denote the beginning and end of the comment/comments.

For example:

```
/*
  Blink
  Turns on an LED for one second, then off for one second,
  repeatedly. This example code is in the public domain.
*/
```

The comment is telling the reader that Blink is the name of the sketch, explains what it does, and advises that the code is in the public domain.

Single-line comments begin with //.

For example:

```
// LED connected to digital pin 13.
int ledPin = 13;
```

Anything after // on any one line is disregarded by the compiler. In this example, "LED connected to digital pin 13" is a comment and is informing anyone who reads it that an LED is connected to digital pin 13.

The second line "int ledPin = 13;" is legitimate code and will be executed by the processor.

Comments are not executable code and are ignored by the processor.

113

Comments can easily be spotted in the Arduino editor as they are highlighted in gray.

There is a naming convention with regard to functions with more than one word in the name. Because there can be no spaces in a name, to make it intelligible, the first letter of each word is capitalized as in "pinMode" opposite.

Setup() and loop() are the most important of all the Arduino's functions. Without either, sketches simply will not run.

Functions

Functions are a set of instructions that can be used anywhere in a sketch. They perform a specific task and, very often, the task is repetitive.

Rather than writing out the code repeatedly, you can use a function to instruct the sketch to perform the task as many times as it is required.

Functions can be either core functions (part of the Arduino language) or ones you have written yourself.

Curly braces are used to denote the beginning and the end of functions. The entire function must be contained within the braces as shown below:

```
void setup() {
   pinMode(led, INPUT);
}
```

Setup() Function

The function shown in the example above is the setup() function and it is an essential part of any Arduino sketch, even if it is not actually used, i.e nothing is entered between the braces.

It is the first function read by a sketch and contains all the instructions necessary to set up the Arduino board. Amongst other things, these include initializing variables, pin modes and libraries, and setting various initial values.

The function only runs once after each power up or reset of the Arduino board.

Loop() Function

Also essential to the running of a sketch, the loop() function is read after the setup() function. Usually containing the majority of a sketch's code, the loop is where the operational functionality of the Arduino takes place.

```
void loop() {
   digitalWrite(13, HIGH);
```

The function is so named because it runs repeatedly, i.e. loops, until something stops it. In so doing, it allows the sketch to change and respond to events such as the inputs from sensors.

Variables

Variables provide a means of naming and storing numerical values so they can be used later on in a sketch. For example, data from a sensor or a value used in a calculation.

As suggested by the name, variables can be changed at any time, unlike constants whose value always remains the same.

Declaring and Initializing Variables

Before they can be used, all variables have to be declared. This means defining the type and, optionally, setting an initial value. The latter is known as initializing the variable.

It is only necessary to declare a variable once. However, even after doing so, the value given to it can always be changed with arithmetic and assignments of various types.

For example:

```
int inputVariable = 0;           // declares a variable & gives it
                                 // a value of 0

inputVariable = analogRead(2);   // Sets the variable to the
                                 // value of analog pin 2
```

The first line of the variable above declares that "inputVariable" is an int (integer), and that it has an initial value of 0. The second line specifies that the variable has the value at analog pin 2.

Once a variable has been declared, it is used by setting the variable equal to the value one wishes to store with it. This is done with the assignment operator, which is an = sign. The assignment operator tells the sketch to put whatever is on the right of it into the variable on the left.

For example:

```
inputVariable = 4;
inputVariable2 = analogRead(2);
```

The first line sets the variable named "inputVariable" to 4. The second line sets "inputVariable2" to the input voltage at analog pin 2.

A variable is basically a label given to a piece of data. It provides a simple method of saving, changing and accessing the data.

115

Variables do not have to be initialized (assigned a value) when they are declared, but it is often useful to do so.

Variables that have to be accessible to more than one function must be set as global variables. This is done by declaring them at the beginning of the sketch.

A variable can be declared in any number of places in a sketch. The location determines which sections of the sketch can use it.

Variables declared inside a loop or function are known as local variables.

...cont'd

Variable Scope

The term "scope" when used in connection with variables in Arduino means the extent to which a given variable is accessible to a particular sketch. This depends on which of the two types of variable is used.

Global Variables

The first type is the global variable. These can be seen and used by every function in a sketch, hence the name. Global variables are declared right at the beginning of a sketch before the setup() function.

For example:

```
int pin = 13;                    // This is a global variable

void setup()
{
   pinMode(pin, OUTPUT);
}
void loop()
{
   digitalWrite(pin, HIGH);
}
```

As you can see, "pin" is used in the setup() *and* loop() functions. As both are referring to the same variable, changing it in one will change it in the other, i.e. the variable is global.

Local Variables

The second type is the local variable. These are declared inside a function and can only be seen and used by that function.

For example:

```
void setup()
{
   int pin = 13;
   pinMode (pin, OUTPUT);
   digitalWrite (pin, HIGH);
}
```

The variable "pin" is only visible to the setup() function. Therefore, it cannot be used by any other part of the sketch.

Variable Data Types

Variable Data Types

By declaring a variable, we are telling the Arduino that we want to save some information in its memory. At the same time, we also have to tell it the data type so it can reserve enough space in the memory.

Computers work in binary – all data is represented by a series of 0s and 1s. These are known as bits. Variables, which are simply bits of information, are comprised of these bits. The more bits used by a variable, the more values, or data, it can contain.

The range of values that can be stored in the Arduino's memory is 2 to the power of the number of bits used. For example, with 8 bits it is possible to store up to 256 different values, and with 16 bits it is possible to store up to 65536 different values.

Variables in Arduino can use 8, 16, or 32 bits. As the Arduino has a limited amount of memory, it is necessary to use the lowest number of bits as possible when coding variables.

To help you do this, the list below shows some common Arduino data types, their bit value, and their purpose:

- **Boolean (8 bit)** - simple logical true/false

- **Byte (8 bit)** - unsigned number from 0 to 255

- **Char (8 bit)** - signed number from -128 to 127

- **Word (16 bit)** - unsigned number from 0 to 6553

- **Int (16 bit)** - signed number from -32768 to 32767

- **Unsigned long (32 bit)** - unsigned number from 0 to 4,294,967,295

- **Long (32 bit)** - signed number from -2,147,483,648 to 2,147,483,647

- **Float (32 bit)** - signed number from -3.4028235E38 to 3.4028235E38

Note that signed variables allow both positive and negative values, while unsigned variables allow only positive values.

Hot tip

Int is the most commonly used data type in Arduino.

Beware

Integer values (int) will roll over if forced past their minimum or maximum values. For example, if X = 32767 and another statement adds 1 to X, X will rollover to -32,768.

Hot tip

Long (32 bit) is an extended data type for long integers with a range of 2,147,483,647 to -2,147,483,648.

Statements

Simple Statements

Statements come in two types – simple and conditional. A simple statement is a single instruction. It takes up a single line and is always followed by a semi-colon.

For example:

```
digitalWrite(ledPin, HIGH);      // This sets ledPin to +5 volts
```

Notice how the statement is followed by a comment that describes what the statement is intended to do.

Conditional Statements

Conditional statements are statements that comprise a condition followed by a series of statements that execute when the condition is met. They enable the flow of a sketch to be controlled in a logical manner.

For example:

```
while (switchValue ==  LOW) {     // This line is the condition
   digitalWrite(ledPin, HIGH);    // This turns the LED on
   delay(1000);                   // Do nothing for 1000 ms
   digitalWrite(ledPin, LOW);     // This turns the LED off
}
```

The first line "while (switchValue == LOW)" is the condition, and is followed by three statements on successive lines. A good way of making sense of conditional statements is to read them logically.

So our example could be read as "while the switch value is low, turn on the LED, do nothing for 1000 milliseconds and then turn off the LED".

Other Arduino conditions include:

- if
- if...else
- for
- do...
- while
- break
- continue
- return

Conditional statements are a very important part of Arduino programming so we'll take a detailed look at them and see how they work.

if – the if statement checks whether a condition has been met, e.g. a value reaching a specific level. If it has, the sketch will run the associated action. If not, the sketch skips the statement and moves on to the next line of code.

The format for an if statement is:

```
if (valueX > valueY) {
   carryOutThisAction;
}
```

In this example, if "valueX" is greater than "valueY", the task specified between the braces is carried out. Note that the symbol > denotes the "greater than" operator.

if/else – the if/else statement is more powerful than the basic if statement, as it allows a number of comparisons to be grouped together.

For example, if you wanted valueX to carry out one action if it was HIGH and carry out a different action if it was LOW, the code would look like:

```
if (valueX == HIGH) {
   carryOutActionA; }
else {
   carryOutActionB
}
```

Note that the symbol == denotes the "equal to" operator.

for – the for statement is useful for situations in which a block of statements needs to be repeated a certain number of times. It allows the block to execute over and over again, i.e. loop, until a specific condition has been met.

Sometimes it is used in conjunction with an increment counter to increment and terminate the loop.

Hot tip

The if statement is the most basic of all the programming statements.

Hot tip

The for statement is often used in combination with arrays.

119

...cont'd

The header for a for loop contains three elements – initialization, condition, and increment or decrement.

For example:

```
for (int x = 0; x > 50; x++) {
    carryOutThisAction;
}
```

The initialization takes place first and occurs just once. It is carried out by the = operator. Next, the condition is tested by the > operator. If it is true, the increment is executed by the ++ compound operator. Lastly, carryOutThisAction is executed.

The whole sequence keeps repeating until such time as the condition is false, at which point the loop ends.

switch/case – the switch/case statement is a more powerful version of the if statement. Assuming three options, you would need three separate if statements – one for each option, or case.

However, just one switch/case statement would be needed as it can switch between the various options.

For example:

```
switch (sensorValue) {
    case 1:
        // do something when sensorValue equals 1
        break;
    case 2:
        // do something when sensorValue equals 2
        break;
    default:
        // if nothing else matches, do the default
        // default is optional
}
```

As you can see, a switch statement compares the value of a variable to the values specified in case statements. When a case statement is found whose value matches that of the variable, the code in that case statement is executed.

While – this statement executes an instruction while a specific condition is true. It will loop, or continue to run, that instruction as long as the condition is true.

As soon as something happens to change the condition to false (the input from a sensor, for example), the loop exits and the instruction is no longer executed.

For example:

```
While (thisValue > 100) {
    carryOutThisAction;
}
```

While "thisValue" is greater than 100, "carryOutThisAction" will be executed continuously. When it falls to below 100, it stops.

do-while – the do-while statement works in the same manner as the while statement, with the exception that the condition is tested at the end rather than at the beginning.

For example:

```
do {
    doSomethingUseful;
} while (x > 50);
```

Here, the code between the curly braces is run before the condition "while (x > 50);" has been evaluated. As a result, even if the condition is not met, the code will run once.

break – the purpose of the break statement is to immediately end a statement and continue the execution of the code that comes after the statement.

For example:

```
while (valueA > 100) {
    if (valueB == HIGH)
{
    break;
}
}
```

The break statement can also be used to separate different sections of a switch/case statement.

...cont'd

continue – the continue statement is used inside loops. It skips over any statements that come after it in a loop and then moves to the next repetition of the loop.

The statement can only be used on repetitive statements such as the while, do and for loops.

For example:

```
for (x = 0; x < 255; x ++) {
    if (x > 40 && x < 120) {
        continue;
    }
    digitalWrite(PWMpin, x);
    delay(50);
}
```

return – the return statement is used to stop the execution of a function and return to the calling function. It does this regardless of where it is. It can also be used to return a value from within a function.

For example:

```
int checkSensor() {
    if (analogRead(0) > 350) {
        return 1;
    else {
        return 0;
}
```

When a return statement is executed, the function is terminated immediately at that point.

Arithmetic & Logic

Arithmetic Operators

Your Arduino is capable of basic binary arithmetic – addition, subtraction, multiplication and division. These operators return the sum, difference, product, or quotient of two operands.

When calculating values, the operations are done with the data type of the operands. This can affect the way the operation is done and thus the result.

For example, when the operands are integers (whole numbers, no decimal points), the results are truncated, not rounded, to the nearest integer. So 13/3 will return 4, not 4.333.

Also, calculations can overflow if the result is larger than the data type is capable of storing. For example, adding 1 to an int with the value 32,767 gives -32,768.

Note that if different types of operand are used, the calculation will use the largest of the types. Also, if your arithmetic requires fractions and thus cannot use integers, you can instead use float variables. However, the drawback is that calculations on floats can be very slow. They can also be large in size and thus use a lot of memory.

The table below shows the arithmetic operators:

Operator	Action
+	addition
−	subtraction
*	multiplication
/	division
%	modulus

These are all self explanatory with the exception of the % modulus operator. This is used to calculate the remainder when one integer is divided by another.

It can be useful in more than one way. One example is to prevent a variable from exceeding a particular range. Another is to detect when a particular value is a multiple of another value.

Hot tip

In mathematics, an operand is the object of a mathematical operation; a quantity on which an operation is performed.

Hot tip

Float is a term used to describe a variable with a fractional value. Numbers created using a float variable declaration will have digits on both sides of a decimal point as opposed to an integer variable, which can only have whole numbers.

Don't forget

Compound operators are useful for keeping your code short and concise.

...cont'd

Compound Operators

Compound operators are used to combine an arithmetic operation with an assignment operation. By doing so, just one variable statement is able to do the same thing as separate arithmetic and assignment operations.

Accordingly, they allow a sketch's code to be written more concisely than would otherwise be the case. Typically, they are used for common operations such as incrementing a value.

For example, to increment value by 1 using arithmetic, you'd write this code:

    anyValue = anyValue +1;

But with a compound operator, you could write it much more concisely:

    anyValue++;

Note that the compound operators ++ and − − only increment or decrement by a value of 1. If you want to use a higher value, you'll need to use the +=, -=,*=, or /= operators.

For example:

    anyValue += 5;

The full list of compound operators is shown in the table below:

Operator	Action
++	increments by 1
− −	decrements by 1
+=	compound addition
− =	compound subtraction
*=	compound multiplication
/=	compound division

Comparison Operators

Comparison operators, as the name implies, allow you to compare two values. This is often done as a means of testing a certain condition to see if it is true or false.

When using comparison operators, make sure all of the operands are of the same data type, i.e. integers should be compared with integers, strings with strings, etc.

In the example below, the if conditional statement is being used in conjunction with the > (greater than) comparison operator:

```
if (anyValue > 20) {
   digitalWrite(LEDpin1, HIGH);
}
```

Hot tip

Don't confuse the assignment operator = (single equal sign) with the comparison operator == (double equal sign).

When the above code is run, the sketch checks to see if "anyValue" is greater than 20. If it is, the sketch carries out the instruction contained with the curly braces, i.e. set the LED connected to pin 1 to HIGH (on). If it isn't, the instruction is ignored.

The list of comparison operators is:

Operator	Action
==	equal to
!=	not equal to
<	less than
>	greater than
<=	less than or equal to
>=	greater than or equal to

Boolean Operators

Boolean, or logical, operators are logic statements that are used to test various conditions in if statements. Each operator will return either a TRUE or a FALSE.

There are three of these operators:

Operator	Action
AND	True only if both operands are true
OR	True if either operand is true
NOT	True if the operand is false

Constants can help to make code easier to read.

Note that the true and false constants are written in lowercase.

Note that true also means any integer which is not 0. So, for example, 1, 56, -300, etc, are all true values.

...cont'd

Note that the words AND, OR and NOT are not written into a sketch's code. Instead they are represented by symbols:

- **AND** – is represented by &&
- **OR** – is represented by ||
- **NOT** – is represented by !

For example:

```
if (x==10 && y==20)
```

The code above will only run if x is 10 and y is 20. If the operator is changed to ||, then it will run if either x is 10 or y is 20. The example below, using the ! operator will only run if x is false.

```
if (!x==0)
```

Constants

A constant is a data value, just as a variable is. The difference between the two is that a constant's value is "read only" so it cannot be changed – hence the name. Constants are declared (written) as "const".

For example:

```
const float pi = 3.14;
```

You can write your own constants or, alternatively, use one of the predefined constants provided by the Arduino programming language.

These are classified in the following three groups:

Boolean Constants – the Boolean constants true and false are used to represent truth and falsity. False is defined as 0 (zero) while true is generally defined as 1.

For example:

```
if (x == true); {
carryOutThisAction;
}
```

HIGH & LOW – when reading or writing to a digital pin there are only two possible values for the pin – HIGH or LOW.

These are set by the HIGH and LOW constants. High is defined as 1, i.e. on (5 volts) while LOW is defined as 0, i.e. off (0 volts).

For example:

  digitalWrite(13, HIGH);

INPUT, OUTPUT & INPUT_PULLUP – the INPUT and OUTPUT constants are used with the pinMode function to set a digital pin as either an INPUT or an OUTPUT.

For example:

  pinMode(13, OUTPUT);

The INPUT_PULLUP constant is a bit more complicated. Pullup resistors are used to ensure that inputs to the Arduino are always at expected logic levels. They do this by ensuring the pins are in either a definite HIGH or LOW state regardless of external factors, e.g. devices being disconnected or the introduction of high impedance values.

Factors such as these can result in the input value at a pin "floating" between HIGH and LOW. LEDs connected to a pin in this state will often flicker faintly on and off. Pullup resistors prevent this.

The microprocessor chip on the Arduino Uno has these pullup resistors built-in. The INPUT_PULLUP constant enables you to activate this option. However, if you prefer, you can use external pullup resistors.

A question often asked by beginners to Arduino is why use a constant when you can use the more flexible variable. One good reason is that a constant makes it absolutely clear to the compiler what it should do.

This allows it to compile the value directly into the sketch. Not only can this eliminate errors, it can also optimize the sketch so it performs more efficiently.

Furthermore, as a bonus, the sketch will use less memory.

Pullup resistors ensure that a pin is at a defined logic level regardless of whether an active device is connected to it.

Pullup resistors can also be used as the interface between different types of logic device.

Arrays

What is an Array

An array is a collection of values that are grouped together and so can be referenced as a whole. They are used when it is necessary to deal with a lot of related data as they help to keep that data organized. Each item in an array is known as an element and the number of elements determines the size of the array.

Arrays can hold anything although, typically, they contain variables. Any type of variable data can be used, e.g. int, char, word, but the data must all be of the same type. When you declare an array, you specify what it will hold. For example:

> int thisArray[] = {1, 2, 3, 4, 5};

This array will hold integer numbers, it is named thisArray, and it has five elements, all of which have been initialized (with the = sign). The first element equals 1, the second equals 2, etc. When the sketch is compiled, the number of elements is counted by the compiler and the array automatically sized accordingly.

However, there are times when the values are not known until a sketch is actually run (inputs from a sensor, for example). So, as you don't know what these values are, you cannot initialize them. In this case, you would declare the array as shown below:

> int thisArray[5];

Here, five elements have been declared (by the number in the square brackets) with the value of each being set to an initial 0. When the sketch is run these values will be updated automatically.

Changing Values

To change the value of an element manually, you enter the number associated with that element between the square brackets. When doing this, be aware that elements are counted from left to right and are numbered from 0.

For example, to change the second element in the array below (currently at 15.9) to 17.5:

> int thisArray[] = {10.5, 15.9, 20.3, 34.7, 42.5};

you'd write:

> int thisArray[1] = 17.5;

Hot tip

Think of an array as a filing box containing a number of index cards, each of which have a piece of related data written on them.

Don't forget

Array elements are numbered from zero. So the first one is 0, the second one is 1, the third is 2, and so on.

Bitwise Operators

In the binary system used by computers, numeric values are written using the numbers 1 and 0. For example, in binary form, the number 667 is written as 1 0 1 0 0 1 1 0 1 1. Each 1 and 0 is known as a bit.

Bitwise operators allow us to perform calculations on integer and byte variables using these binary representations. The main advantages of doing this are that sketch speed is improved, and that less of the Arduino's memory is needed.

To demonstrate how this works, we'll look at the Bitwise AND operator. This is written as &, and the operator is used to compare two binary numbers at a bit-by-bit level.

The syntax for this is:

   byte a & b

An example calculation is:

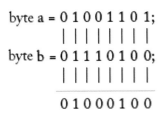

```
byte a = 0 1 0 0 1 1 0 1;
         | | | | | | | |
byte b = 0 1 1 1 0 1 0 0;
         | | | | | | | |
         _____
         0 1 0 0 0 1 0 0
```

If the bits in any one column are both 1, the output is 1. If not, the output is 0.

The Bitwise operators are:

- **AND (&)** – outputs a 1 if if both values in a column are 1

- **OR (|)** – outputs a 1 if either value in a column is 1

- **XOR (^)** – outputs a 1 if the bits in a column are different

- **NOT (~)** – changes 0s to 1, and 1s to 0

- **Left (<<)** – moves bits to the left by a specified amount

- **Right (>>)** – moves bits to the right by a specified amount

Our example AND calculation on the left is done on two byte values, a and b. As we saw on page 117, byte data types are 8-bit, meaning there are 8 bits in the number. Therefore, the calculation involves 8 simultaneous operations, one for each bit.

A common use of Bitwise AND is to access specific bits in a byte of data. This is known as "masking".

Input & Output Interfaces

By itself, your Arduino is very limited in what it can do. Connect it to an external circuit, however, and there is little it cannot do. This requires a means of connection, or an interface as it is known. The Arduino provides two types:

- Digital pins (used for digital signals)

- Analog pins (used for analog signals)

These pins can be configured either as inputs or as outputs. Also, some of the analog pins can be converted to digital.

To let you do this, the Arduino provides a number of functions with which to configure the input/output pins to suit your purposes.

We'll look at configuring digital pins first.

Digital Inputs & Outputs

On pages 20-21, we saw that the Arduino Uno provides 14 digital pins (interfaces). These can be configured with the following functions:

pinMode() – the pinMode() function is used to set a digital pin as either an input or an output. The syntax is simple as we see in the example below:

```
pinMode(pin, mode);
```

The function has two parameters – pin and mode. "pin" is the number of the pin being set up and can be anywhere between 0 to 13. The "mode" parameter determines whether the pin functions as an INPUT, an INPUT_PULLUP or an OUTPUT.

The INPUT constant sets the pin to a high impedance state whereby the pin makes very little demands on the load circuit. This is advantageous in low current circuits, such as those incorporating sensors.

The OUTPUT constant, on the other hand, sets the pin to a state of low impedance. This enables the pin to provide a substantial amount of current to other circuits.

The INPUT_PULLUP constant sets the pin as an INPUT and also activates the pin's internal pullup resistor.

Hot tip

Digital pins, by default, are set as INPUTS. This means that they don't have to be specified as such.

digitalWrite() – when a digital pin has been set to OUTPUT mode, it then needs to be set as either on or off. This is done with the digitalWrite() function, the syntax for which is:

```
digitalWrite(pin, mode);
```

As with pinMode(), the function has two parameters: "pin" refers to the pin being used and "mode" refers to one of two constants – HIGH or LOW.

If you want to set a pin, say pin 10, to on, the code will be:

```
digitalWrite(10, HIGH);
```

The HIGH constant instructs the Arduino to send 5 volts to the specified pin at a current of about 40mA. The LOW constant sets the pin to 0 volts emulating, in effect, a ground connection.

digitalRead() – when a digital pin has been set to INPUT, the state of the pin can be determined with the digitalRead() function as shown below:

```
digitalRead(pin);
```

The output of the function is either HIGH or LOW and can be used in two ways.

The first is to simply assign the state of the pin to a variable. The second is to use the function as a variable. Consider the following example code:

```
someVariable = digitalRead(pin2);
if (someVariable == LOW) digitalWrite(ledPin, LOW);
```

In the first line, the digitalRead() function reads the value at INPUT pin (pin2) and then assigns it to a variable (someVariable). In the second line, a conditional if test is done on the variable and if it is equal to LOW, it sets an output pin (ledPin) as LOW.

Analog Inputs & Outputs
Not all signals are digital of course. The output of many devices, motors and sensors for example, are in analog form. As the Arduino is a digital device, it has to convert these analog signals to digital form before it can use them.

Before you can use the digitalWrite() function to turn a pin on or off, you must first use the pinMode() function to specify the pin as an INPUT or OUTPUT.

...cont'd

Conversely, sometimes it is necessary to convert digital signals from the Arduino to analog form that can be used by external analog devices.

The Arduino Leonardo is the only Arduino model that comes with a digital-to-analog converter. All the other models use the PWM method.

To convert analog to digital, the Arduino has an analog-to-digital converter (ADC). To convert digital to analog, it uses the pulse width modulation (PWM) technique. As we have already seen, this is a simulation achieved by controlling an analog signal's duty cycle.

The two functions used to configure analog pins are:

analogRead() – this function produces an integer value that represents the analog signal at the pin. This value ranges from 0 to 1023.

For example:

    value = analogRead(A2)

On the Arduino Uno, PWM is only supported on pins 3, 5, 6, 9, 10 and 11.

The Arduino's analog pins are prefixed A0 to A5. Thus the "pin" will be within this range – A2 as shown above. The "value" is set at the same value as at pin A2.

analogWrite() – when you need an analog output signal at a certain pin, you have to use the analogWrite() function. As we mentioned above, this employs the Arduino's PWM feature to simulate the required signal.

The syntax is:

    analogWrite(pin, someValue)

The function writes the value at "someValue" to the specified pin.

A value of 0 generates 0 volts at the specified pin and a value of 255 generates 5 volts.

Time

When running applications that operate in real time, your sketches will usually need to be aware of time and time-related concepts.

The Arduino has this covered with four related functions as we see below:

delay() – the delay() function pauses a sketch, or waits for something to happen, for the number of milliseconds specified in the code.

For example:

delay(1000)

The figure entered between the parentheses determines the delay – 1000ms, or 1 second.

delayMicroseconds() – this is the same as the delay() function but the delay period is in microseconds.

Note that the largest value that will produce an accurate delay is 16383. For delays longer than a few thousand microseconds, it is recommended that you use the delay() function instead as it will be more accurate.

Both of these functions are useful in situations where you need a sketch to wait for a specified period, or where you just need to slow things down a bit.

millis() – the millis() function tells you the length of time in milliseconds that the current sketch has been running.

micros() – this function is the same as the millis() function but returns the required value in microseconds rather than in milliseconds.

Apart from telling you how long a particular sketch has been running, these two functions give you a sense of time scale in a sketch.

Hot tip

In both the millis() and micros() functions, the number will overflow, i.e. be reset to zero, after a certain period.

Other Useful Functions

Your Arduino offers a number of functions other than the ones mentioned in this chapter so far. These include:

Advanced Arithmetic

For most sketches, the basic arithmetic operators (+, -, x, /) are usually all you need. More complex sketches, though, may require something more advanced. For these cases, Arduino provides the following arithmetic and trigonometry functions:

- **min(x, y)** – returns x or y, whichever is the lower value

- **max(x, y)** – returns x or y, whichever is the highest value

- **abs(x)** – returns the absolute value of x

- **constrain(x, a, b)** – constrains a number within a range. x is returned if it is between a and b, a is returned if x is lower than a, and b is returned if x is higher than b

- **map(x)** – re-maps the value x from one range to another

- **pow(x, y)** – returns the value of x raised to the power of y

- **sqrt(x)** – returns the square root of value x

- **sin(x)** – returns the sine of value x

- **cos(x)** – returns the cosine of value x

- **tan(x)** – returns the tangent of value x

Random Numbers

Many computing applications require random numbers. Common examples are password and encryption key generation, digital games, shuffling audio files, etc.

The majority of programming languages provide a random number generating function and the Arduino language is no exception (although the numbers generated are pseudo-random).

It provides two functions for this purpose:

random(min, max) – the random() function returns random numbers within a range specified by min and max values. Note that the min value is optional – if you don't specify it, it defaults to 0.

Beware

Don't use other functions with abs(). Because of the way it is implemented, doing so can cause errors.

Hot tip

map(x) and constrain() are commonly used with sensors as they enable sensor outputs to be kept within a specific range.

For example:

```
valueA = random(50, 100);
```

This sets the value at valueA to a random number within the range 50-100.

randomSeed(seed) – the randomSeed() function initializes the pseudo-random number generator, causing it to start at an arbitrary point in its random sequence.

Seed values are integers that define the exact sequence of the pseudo-random numbers generated. Even the tiniest change in seed value will result in a radically different random sequence.

Bit Level Manipulation

Bit manipulation is a process in which bits of data shorter than a word are manipulated using algorithms. Basically, it allows data to be altered at bit level.

This process is often needed with programming tasks such as error detection, data compression, encryption, and program optimization.

The Arduino programming language provides a range of bit manipulation functions.

These are:

- **lowByte()** – returns the low-order byte of a word

- **highByte()** – returns the high-order byte of a word

- **bitRead()** – reads a bit of a number

- **bitWrite()** – writes a bit of a numeric value

- **bitSet()** – sets (writes a 1 to) a bit of a numeric value

- **bitClear()** – clears (writes a 0 to) a bit of a numeric value

- **bit()** – returns the value of the specified bit

Hot tip

Computer random number generators tend to generate repeatable random sequences rather than truly random numbers. The randomSeed() function alleviates this limitation to a certain degree by enabling a variable, constant or other function to be inserted into the random function.

135

Hot tip

A bit is the basic unit of information in computing. It has a value of either 1 or 0. These values can also be interpreted as TRUE/FALSE, HIGH/LOW, ON/OFF, etc.

Sketch Structure

In this chapter, we have seen that the basic elements used in writing Arduino sketches are variables, loops, statements, input/output, and functions. The ones actually used vary from sketch to sketch but the one thing that doesn't vary is the basic sketch structure – all sketches consist of four main sections. These are:

- Comments (optional)

- Global variables (if any)

- Setup()

- Loop()

This is demonstrated in the sketch code shown below:

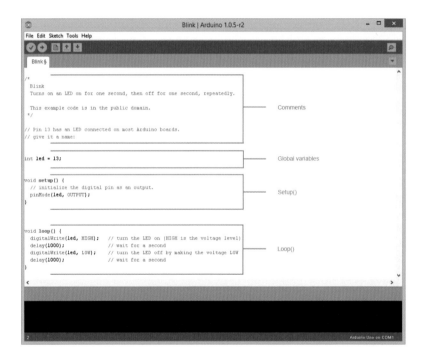

Of these, comments are entirely optional. Variables can be placed anywhere but if they need to be accessible to the entire sketch, rather than just a part of it, they must go at the top below the general comments.

The setup() and loop() functions are the only part of a sketch that are essential. If either is not there, the sketch will not run.

9 Sketches

We begin this chapter by writing the code for a simple sketch. We then analyze a number of the Arduino's in-built sketches seeing what hardware is required, how they work and what they do.

Write an Arduino Sketch

You've learned about the various elements used in Arduino sketches – comments, functions, statements, loops, etc. You know about syntax and how important it is to get it right. Now it's time to put it all together and write your first sketch.

Basics

The first step is to connect the Arduino to your computer and open the Arduino IDE. The editor window automatically opens a new sketch. The name of the sketch will be in the tab at the top-left of the editor window in the following format:

sketch_mmmddx

mmm is the first three letters of the month, dd is the date, and x is a letter that differentiates between sketches created on the same day.

When you begin typing code in the editor window, you'll notice that specific sections of the code will be highlighted in different colors. For example, function keywords are orange and constants are blue. This makes it easier to identify syntax errors and thus eliminate mistakes that will prevent your sketch from compiling.

Comments

Before you start entering the code, you may want to make some comments with regard to the sketch. For example, you can give the sketch a name, say when it was written, by whom it was written and what it is intended to do.

As you progress through the sketch you can add comments at any point to help others understand what a particular section or line of code does, as reminders to yourself, etc. These can prove to be extremely useful both to you, and to others.

As we saw on page 113, single-line comments are preceded by two forward slashes:

// Blinking LED sketch by John Smith, written on 12/11/2014

and multi-line comments are preceded with /* and ended with */:

```
/*
Blinking LED sketch by John Smith
written on 12/11/2014
*/
```

Hot tip

The "Blinking LED" sketch we show how to write on pages 138-141 is a variation of the Blink example sketch available in the Arduino IDE.

Write your own version of the comments shown on page 138, click the Save icon, give your sketch a name and then click OK

If you now go to **File > Sketchbook**, you'll see your sketch is listed there.

The Setup Function

Next, you need to set things up with the setup() function. This will be the first section of executable code in the sketch and, for this reason, it contains all the instructions needed to set up the board. For example, initializing variables, setting pin modes, and calling libraries.

These instructions are entered between the braces and run just once. The syntax for the setup() function is:

```
void setup() {
}
```

The purpose of the sketch we are writing is to make the LED on the Arduino blink on and off. As the LED is connected to pin 13 by default, the setup instructions will tell the Arduino to set pin 13 as an output. It will then send 5 volts to the LED thus turning it on.

 Enter the following code in the editor after the comments

```
void setup() {
   pinMode(13, OUTPUT);
}
```

pinMode is the function that sets pin 13 as an output (if we were to write INPUT instead, it would be an input). The sketch so far will now look something like this:

```
/*
Blinking LED sketch by John Smith
written on 12/11/2014
*/

void setup() {
   pinMode(13, OUTPUT);
}
```

Save your work periodically, particularly when writing long or complex sketches.

It is not essential to place instructions in the setup() function. Some simple sketches may not actually need any. However, the function itself has to be present even if nothing is put between the braces.

Code in the setup function basically gets the Arduino ready for action, and the code in the loop function puts it into action.

...cont'd

The Loop Function

Next is the loop function. Typically, this is where the majority of a sketch's code is placed.

Whereas the setup function initializes the Arduino's input and output pins and gets them ready, the loop function puts them into action.

Loop function code is executed repeatedly until it is stopped by the power being removed or the reset button on the board being pressed. The syntax for the loop function is:

```
void loop() {
{
```

1 As with the setup function, the loop code is placed within the braces. So, continuing with our Blinking LED sketch, enter the following code after the setup function code

```
void loop() {
   digitalWrite(13, HIGH);
   delay(2000);
   digitalWrite(13, LOW);
   delay(2000);
}
```

In explanation, the digitalWrite function controls the voltage that is sent from a digital pin (pin 13 in our case).

It does this by setting the value at the pin to either HIGH, in which case 5 volts is sent from the pin thus lighting the LED; or LOW, in which case there is no voltage at the pin, so the LED is off.

The loop also contains two delay() functions. These tell the sketch to stop running for the specified period – 2000ms, or 2 seconds.

The sequence in our loop is:

● The first digitalWrite function sends 5 volts to pin 13 thus turning the LED on.

● The first delay function stops everything for 2 seconds, so the LED remains on for 2 seconds.

- The first delay function ends allowing the second digitalWrite function to run. This removes the power from pin 13 thus turning the LED off.

- The second delay function stops everything for 2 seconds again thus keeping the LED off for 2 seconds.

The loop then repeats itself and will continue to do so until it is stopped.

The complete sketch will look like this:

```
/*
Blinking LED sketch by John Smith
written on 12/11/2014
*/

void setup() {
   pinMode(13, OUTPUT);
}

void loop() {
   digitalWrite(13, HIGH);
   delay(2000);
   digitalWrite(13, LOW);
   delay(2000);
}
```

Note that in our Blinking LED sketch, there are no global variables. If there were (as there are in most sketches), they would be entered after the comments but before the setup() function.

Don't forget

Global variables (as opposed to local variables) must go before the setup function. Library calls go right at the top of the sketch.

Verify the Sketch

Also known as compiling, verification of a sketch is the Arduino checking that the syntax is correct, i.e. that it has been written in the format that the Arduino understands. Do it as follows:

1 When you have finished writing the sketch code, click the **Verify** icon at the far-left of the IDE

2 After a few seconds (depending on the size of the sketch), you should see the following message

3 The message says "Done compiling" and shows you the size of the sketch in bytes that will be uploaded to the Arduino. This tells you the sketch is formatted correctly and should load without any issues

4 If, however, if you get an error message as shown below, then you need to correct the error before the verification procedure will complete – see pages 170-174

5 The message will tell you which line contains the error and also the nature of the error

Hot tip

A sketch will not compile if it contains *any* syntax errors.

Hot tip

The size of a sketch can be important as the Arduino has a limited amount of memory. The verification procedure tells you what it is.

Upload the Sketch

Verification done, you are now ready to upload your sketch.

1 Assuming there is a good connection between the board and your computer, simply click the **Upload** button in the Arduino editor window

If your sketch won't upload, go to **Tools > Board** and make sure Arduino Uno (or whatever board you are using) is selected.

2 The sketch will be compiled again by the IDE and then uploaded. During this procedure you will see the TX/RX LEDs on the Arduino flashing – this indicates data is flowing to and from the board

3 If all goes to plan, the sketch will now run. You don't need to press a Run button as such – as soon as the upload is complete it runs automatically. Look at the board and you will see the LED blinking on and off

Sketch Modifications

You may want to edit your sketch at some point to change the way it works. The Arduino IDE makes this easy. Click the **Open** button and, assuming you've previously saved the sketch, you'll see it at the top of the sketch list. Click to open it. Make the required change, click **Save** and then click **Upload**.

An upload keyboard shortcut for Windows is Ctrl+U. For Mac OS X, try Cmd+U.

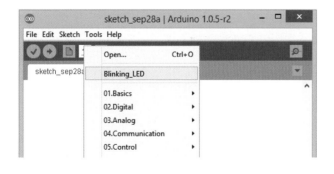

The Fade Sketch

Our Blinking LED sketch is a very simple example intended to familiarize you with the essential parts of a sketch, and the procedures of verifying and uploading it. Let's now take a look at something more complex – the Fade Sketch.

This demonstrates how the analogWrite() function can be used to fade an LED on and off by utilizing the pulse width modulation (PWM) capabilities of the Arduino board. As we have already seen, PWM provides a method of converting a digital signal to an analog signal. The analog signal is then used to progressively vary the brightness of the LED, thus gradually fading it on and off.

The Circuit

To build the circuit you will need a 220 ohm resistor and a light emitting diode (LED) of any color.

A true digital signal is a square wave and can thus only ever be on or off. Therefore, if used to power an LED, the LED would also be either on or off with no variation in brightness in between.
  The analogWrite() function provides a way of simulating an analog signal at a digital pin.

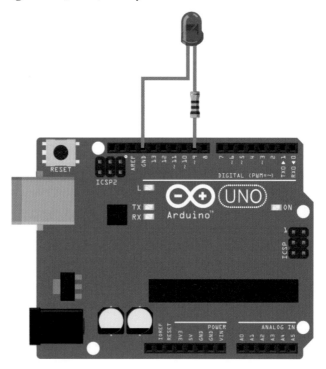

A 220 ohm resistor has color bands of red, red and brown.

144

1 Connect the short leg of the LED to the GND pin and connect the other leg to pin 9 via the resistor

2 Now, connect the board to your PC with the USB cable and open the Arduino IDE

…cont'd

3 Go to **File > Examples > 01.Basics > Fade**

4 Click the **Upload** button and then take a look at the Arduino board. You will see the brightness of the LED fade in and out

In the editor, you will see the following code:

```
/*
  Fade

  This example shows how to fade an LED on pin 9
  using the analogWrite() function.

  This example code is in the public domain.
  */

int led = 9;           // the pin that the LED is attached to
int brightness = 0;    // how bright the LED is
int fadeAmount = 5;    // how many points to fade the LED by

// the setup routine runs once when you press reset:
void setup() {
  // declare pin 9 to be an output:
  pinMode(led, OUTPUT);
}

// the loop routine runs over and over again forever:
void loop() {
  // set the brightness of pin 9:
  analogWrite(led, brightness);

  // change the brightness for next time through the loop:
  brightness = brightness + fadeAmount;

  // reverse the direction of the fading at the ends of the fade:
  if (brightness == 0 || brightness == 255) {
  fadeAmount = -fadeAmount ;
  }
  // wait for 30 milliseconds to see the dimming effect
  delay(30);
}
```

...cont'd

Sketch Analysis

At the beginning of the sketch, we have a multi-line, or block, comment describing the purpose of the sketch.

```
/*
   Fade

   This example shows how to fade an LED on pin 9
   using the analogWrite() function.

   This example code is in the public domain.
   */
```

This is followed by a section of code that declares and initializes three int global variables:

```
int led = 9;             // the pin that the LED is attached to
int brightness = 0;      // how bright the LED is
int fadeAmount = 5;      // how many points to fade the LED by
```

The led variable states which pin the LED is attached to. The brightness variable stores the current brightness of the LED, while the fadeAmount variable determines the rate of the LED's change in brightness.

The next section of code is the setup() function:

```
// the setup routine runs once when you press reset:
void setup() {
   // declare pin 9 to be an output:
   pinMode(led, OUTPUT);
}
```

This does just one thing: it uses the pinMode() local function to set pin 9 as an output.

Next, we have the loop function:

```
// the loop routine runs over and over again forever:
void loop() {
   // set the brightness of pin 9:
   analogWrite(led, brightness);

   // change the brightness for next time through the loop:
   brightness = brightness + fadeAmount;
```

Hot tip

The pinMode function uses two values: The first is the pin number (1, 2, 3, etc.) and the second is the mode (INPUT or OUTPUT).

Don't forget

The loop() function usually contains the main part of a project's code.

```
// reverse the direction of the fading at the ends of the fade:
if (brightness == 0 || brightness == 255) {
fadeAmount = -fadeAmount ;
}
// wait for 30 milliseconds to see the dimming effect
delay(30);
}
```

The first piece of code here is "analogWrite(led, brightness);" This activates the Arduino's pulse width modulation (PWM) feature, which lets us vary the power at the output (pin 9) by turning the digital signal into an analog one. This in turn allows the brightness of the LED at pin 9 to be varied, instead of being either just on or off, as it would be with a square wave digital signal.

To explain this in more detail:

The "analogWrite(led, brightness)" function contains two arguments – led and brightness. Referring back to the three variables at the beginning of the sketch, we see that in the first one, "int led = 9", we have declared that the led is attached to pin 9. In the second, "int brightness = 0", we have declared that the brightness of the LED is 0, or dark. The "analogWrite(led, brightness)" function calls the values set in these two variables.

Next, is "brightness = brightness + fadeAmount;" Going back to the variables, this function takes the current brightness of the LED, which is 0, and then adds the fadeAmount, which is 5. Thus we have a total value of 5, which is saved back to the int brightness variable.

So, whereas initially it is 0 and the LED is thus dark the first time the loop runs, the next time it will be 5 and the LED will begin to glow. As the code is repeated in the loop, the value saved to the int brightness variable rapidly increases. Each time this happens, the power to the LED increases making it glow more brightly.

Eventually the analogWrite() limit of 255 is reached and the LED won't glow any brighter. Thus a way is needed to ascertain the value of the brightness variable and then change it when it gets to the limit.

This introduces the next section of code – the if statement.

Hot tip

The amount of power provided by PWM varies on a scale of 0 to 255.

Hot tip

Some important points with regard to the analogWrite() function include:

It has nothing to do with the analog pins.

It works with pins 3, 5, 6, 10 and 11.

It uses the PWM feature to provide an adjustable power at the output pin.

...cont'd

This checks a condition and, if it is met, carries out an action. If it isn't met, it does nothing.

```
if (brightness == 0 || brightness == 255) {
fadeAmount = -fadeAmount;
```

The condition to be checked (brightness) is placed inside the brackets ( ), and the action to be carried out (fadeAmount) if it is met is placed inside the braces { }. if statements are either TRUE, in which case the code within the braces is executed; or FALSE, in which case it is not executed.

Don't get confused between the = and == symbols. The former is used to assign a value whereas the letter is a comparison operator.

Looking at the first line, we see the || symbol. This means OR. Therefore the condition is stating that "if the brightness variable is equal to zero OR the brightness variable is equal to 255".

You'll also notice the == symbol. This is a comparison operator that asks "are these values the same?"

In our Fade sketch, the brightness value has reached the maximum of 255 so the "if (brightness == 0 || brightness == 255)" condition has been met. So it carries out an action and this action is the "fadeAmount = -fadeAmount;" variable.

All that happens in the action is that a negative sign (-) is placed before the fadeAmount value thus turning it from a positive value to a negative one. Each time this code is repeated in the loop, a value of 5 is taken from the brightness variable, gradually reducing the power at the LED and thus its brightness.

When the brightness variable reaches 0, the fadeAmount action is switched to positive and the whole process begins again.

Finally, the speed at which the fade process occurs needs to be reduced to one that is discernible. If we didn't do this, the LED would appear as a steady unchanging light. This is done with the last bit of code:

```
// wait for 30 milliseconds to see the dimming effect
delay(30);
}
```

The delay function simply applies a delay of 30 milliseconds, which is enough to make the change in the LED's brightness discernible to the human eye.

The DigitalReadSerial Sketch

This sketch allows you to monitor the state of a switch or sensor, i.e. is it on or off? This enables you to initiate actions accordingly – if the button is depressed, carry out action A, if it isn't, carry out action B.

To be able to do this, it is necessary to see the status of the Arduino's pins and track any changes as they occur. The serial monitor provides us with the means of doing it.

The Circuit
The following parts will be required:

- Push button
- 10,000 ohm (10K) resistor
- Jumpers
- Breadboard

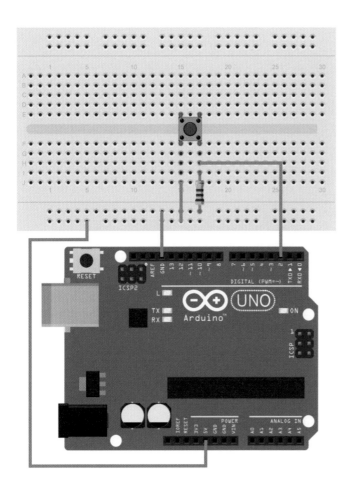

...cont'd

Fit the switch on the breadboard as shown. Then connect one side of it to the 5 volt pin on the Arduino board, and connect the other side to the GND pin via the 10K resistor. Also connect this side to pin 2 on the Arduino board.

Connect the board to the computer with the USB cable and open the IDE. Go to **File > Examples > 01.Basics** and click **digitalReadSerial**. Then click the **Upload** button.

When the sketch has loaded, open **Tools > Serial Monitor** on the menu bar. The serial monitor window will open and you will see a series of 0s. Press the button and keep it depressed – a series of 1s will be displayed.

In the editor, you will see the following code:

Hot tip

The Arduino continuously reads the state of pin 2 to see if the button is open or closed.

When it is open (unpressed), pin 2 is connected to ground through the resistor and thus reads LOW, or 0. When it is closed (pressed), it connects pin 2 to 5 volts, thus the pin reads HIGH, or 1.

```
/*
  DigitalReadSerial
  Reads a digital input on pin 2, prints the result to the serial
  monitor

  This example code is in the public domain.
*/

// digital pin 2 has a push button attached to it. Give it a name:
int pushButton = 2;

// the setup routine runs once when you press reset:
void setup() {
  // initialize serial communication at 9600 bits per second:
  Serial.begin(9600);
  // make the push button's pin an input:
  pinMode(pushButton, INPUT);
}

// the loop routine runs over and over again forever:
void loop() {
  // read the input pin:
  int buttonState = digitalRead(pushButton)
  // print out the state of the button:
  Serial.println(buttonState);
  delay(1);        // delay in between reads for stability
}
```

Sketch Analysis

The sketch begins with a block comment that describes the purpose of the sketch, i.e. read the value at pin 2 and print it to the serial monitor.

This is followed by the section of code that declares and initializes the sketch's variables. In this sketch, there is just one:

```
// digital pin 2 has a push button attached to it. Give it a name:
int pushButton = 2;
```

The comment tells you all you need to know – the push button on the circuit is connected to pin 2.

Next, we have the setup() code:

```
// the setup routine runs once when you press reset:
void setup() {
  // initialize serial communication at 9600 bits per second:
  Serial.begin(9600);
  // make the push button's pin an input:
  pinMode(pushButton, INPUT);
}
```

This contains two functions: Serial.begin() and pinMode(). Serial.begin() belongs to a related group of library functions – in this case the Serial library. The word Serial is the name of the library and the word after the fullstop is the name of the function.

So "Serial.begin(9600);" is calling the begin function in the Serial library. This activates the serial monitor, which allows you to see exactly what's happening on the Arduino in a monitor window on the computer.

Following Serial.begin(9600) is the pinMode() function. This enables a pin to be set either as an INPUT or an OUTPUT. In this case, the word "pushbutton" calls the value specified in the "int pushbutton = 2" variable at the beginning of the sketch.

Thus pin 2 is the selected pin and it is set as an INPUT.

Next, is the loop function:

Hot tip

A library is a group of functions that, together, do something specific. Your Arduino comes with a number of built-in libraries (including the Serial library). You can also download and use community libraries.

...cont'd

```
// the loop routine runs over and over again forever:
void loop() {
    // read the input pin:
    int buttonState = digitalRead(pushButton)
    // print out the state of the button:
    Serial.println(buttonState);
    delay(1);        // delay in between reads for stability
}
```

The first line of code is the following variable:

```
int buttonState = digitalRead(pushButton)
```

This declares an integer value and names that value buttonState. The purpose of the variable is to hold the state of the button, i.e. is it on or off?

The variable is initialized by setting it as equal to the output of the digitalRead() function. This function returns a value of either 1 or 0, with 1 indicating the voltage at the pin is high and 0 indicating it is low. The pin in question, of course, is pin 2. This is because pushButton is the value specified in the variable's brackets and pushButton has already been initialized as pin 2 in the global variable at the beginning of the sketch.

As a result, when the push button is pressed it completes the circuit between pin 2 and the 5 volt output from the Arduino. This is read by the digitalRead(pushbutton) function and interpreted as a value of HIGH. When the button is not pressed, pin 2 is connected to GND so has no voltage. This is read and interpreted as a value of LOW.

Next is the Serial.println() function:

```
Serial.println(buttonState);
```

The Serial.println() function is taken from the same Serial library as the Serial.begin() function. It reads the state of the button and prints it to the serial monitor in the form of 1s and 0s.

Lastly, the delay() function slows things down so the data can be read more easily, in this case by 1 millisecond.

```
    delay(1);        // delay in between reads for stability
}
```

Hot tip

Don't be confused by the Serial.println() function – it has nothing to do with your paper printer. All it does is send the ouput of the Arduino to a monitor on your computer.

The AnalogReadSerial Sketch

The digitalReadSerial sketch we've just looked at enables us to ascertain digital values, i.e. if something is either on or off. This is fine as far as it goes but often we need to know more than this.

Take, for example, an audio sensor that provides information such as pitch and amplitude. Sensors of this type return analog values that can cover a large range – not just a simple on or off.

Enter the AnalogReadSerial sketch. This piece of code reads the signals at your Arduino's analog input pins – A0 to A5.

The Circuit
To build the required circuit, you will need the following parts:

● Potentiometer
● Breadboard
● Jumpers

The Arduino's analog inputs can produce a value anywhere between 0 and 1023. 0 equates to 0V, 512 to 2.5V, and 1023 to 5V.

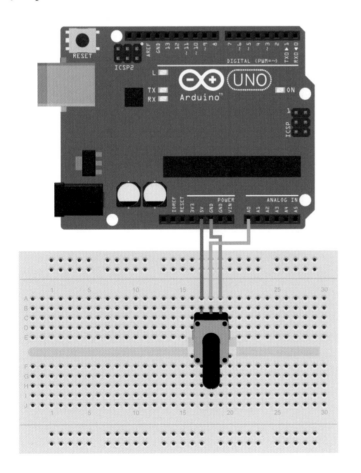

Hot tip

The analog input pins are connected to an analog-to-digital converter. This takes the analog input signals and turns them into digital versions that the Arduino can understand.

153

...cont'd

Fit the potentiometer on the breadboard as shown. Then connect one of the potentiometer's two outside leads to the 5 volt pin on the Arduino board and the other one to GND. Finally, connect the potentiometer's middle lead to pin A0 on the Arduino board.

Connect the Arduino to the computer with the USB cable, open the Arduino IDE and go to **File > Examples > 01.Basics > AnalogReadSerial**.

Click the **Upload** button.

Now, go to **Tools > Serial Monitor**. This will open the serial monitor window, which displays the information being read from the Arduino – in this case a range of numbers between 0 and 1023. Turning the potentiometer's knob will cause these numbers to change.

In the editor window, you will see the following code:

Hot tip

We are using a potentiometer in this sketch as it produces an output that varies. You could instead use any type of sensor.

```
/*
  AnalogReadSerial
  Reads an analog input on pin 0, prints the result to the serial
  monitor.
  Attach the center pin of a potentiometer to pin A0, and the
  outside pins to +5V and ground.

  This example code is in the public domain.
*/

// the setup routine runs once when you press reset:
void setup() {
  // initialize serial communication at 9600 bits per second:
  Serial.begin(9600);
}

// the loop routine runs over and over again forever:
void loop() {
  // read the input on analog pin 0:
  int sensorValue = analogRead(A0);
  // print out the value you read:
  Serial.println(sensorValue);
  delay(1);        // delay in between reads for stability
}
```

Sketch Analysis

As always, the sketch begins with a block comment that describes what the sketch is intended to do.

Note that no variables are declared after the initial block comment as is usual in most sketches. The reason for this is that in this particular sketch, the variable is declared inside the loop function.

Following the block comment is the setup() function.

```
    // the setup routine runs once when you press reset:
void setup() {
    // initialize serial communication at 9600 bits per second:
    Serial.begin(9600);
}
```

All we have in setup() is the Serial.begin() function that initiates the serial monitor at a speed of 9600 baud. As we have already seen, the serial monitor is a communication channel that connects the computer to the Arduino board and enables you to see exactly what is happening on the board.

Note that in this sketch we haven't used the pinMode function to set pin A0 as an input. The reason is that it isn't necessary as, by default, the Arduino automatically sets all pins as inputs.

However, for reasons of clarity, it is usually advisable to do so. In our sketch, the code "pinMode(A10, INPUT)" would go between the braces in the setup() function.

Next, is the loop() function:

```
    // the loop routine runs over and over again forever:
void loop() {
    // read the input on analog pin 0:
    int sensorValue = analogRead(A0);
    // print out the value you read:
    Serial.println(sensorValue);
    delay(1);       // delay in between reads for stability
}
```

Pins on the Arduino board do not have to be specifically set as INPUTS. This is done by default.

...cont'd

As with the digitalReadSerial sketch, the loop function begins with the declaration and initialization of a variable:

```
int sensorValue = analogRead(A0);
```

This is given the name sensorValue and it is initialized by setting it as equal to the output of the analogRead() function.

Hot tip

The value produced by an analog input is proportional to the voltage at the input.

The analogRead function reads the value at a specified pin – pin A0 in our sketch. This pin is connected to the potentiometer's middle pin, the voltage at which is mapped to a number anywhere between 0 and 1023. This is done by the digital-to-analog converter. The number, or value, is then assigned to the sensorValue variable.

Note that the actual voltage at pin A0 is between 0 volts and 5 volts, depending on the position of the potentiometer's adjustment knob.

The loop also includes the Serial.println() function. This activates the serial monitor with which it is possible to view the values produced by the sketch at pin A0, as shown below:

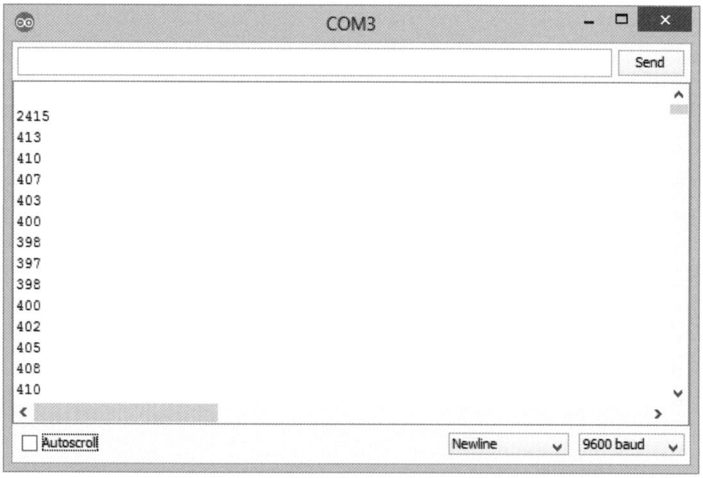

Hot tip

If you find the values are scrolling down too fast, set a longer delay in the delay() function.

If you adjust the potentiometer you will see the values change in the monitor window.

The delay(1) function introduces a delay of 1 millisecond between reads to slow things down making it easier to read.

The IfStatementConditional Sketch

The Fade sketch demonstrated how the if statement is used to perform an action if a certain condition has been met. In the ifStatementConditional sketch, we see how to stack different conditions to provide flexibility in your sketches.

The Circuit
To build the circuit, you will need the following:

- Potentiometer
- 220 ohm resistor
- LED
- Breadboard
- Jumpers

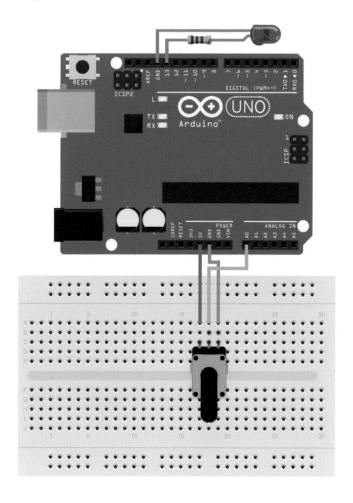

...cont'd

Fit the potentiometer on the breadboard as shown. Connect either of its outside leads to the 5 volt pin on the board and the other lead to a GND pin. Then connect its middle lead to the A0 pin. Connect the LED and resistor as you did in the Fade sketch but this time use pin 13 instead of pin 9.

Now connect the Arduino to your PC with the USB cable and open the Arduino IDE. Go to **File > Examples > 05.Control > ifStatementConditional**.

Click the **Upload** button.

You will see the following code in the editor:

Hot tip

Note that in the sketch code on the right, we have not included the comment block at the beginning of the sketch.

```
// These constants won't change:

const int analogPin = A0;      // pin that the sensor is attached to
const int ledPin = 13;         // pin that the LED is attached to
const int threshold = 400;     // an arbitrary threshold level that's in
                               //   the range of the analog input

void setup() {
    // initialize the LED pin as an output:
    pinMode(ledPin, OUTPUT);
    // initialize serial communications:
    Serial.begin(9600);
}

void loop() {
    // read the value of the potentiometer:
    int analogValue = analogRead(analogPin);

    // if the analog value is high enough, turn on the LED:
    if (analogValue > threshold) {
    digitalWrite(ledPin, HIGH);
    }

    else {
      digitalWrite(ledPin,LOW);
    }

    // print the analog value:
    Serial.println(analogValue);
    delay(1);        // delay in between reads for stability
}
```

Sketch Analysis

The first thing to understand here is how the potentiometer works. At one terminal it has 5 volts and at the other it has 0 volts. As the knob is rotated backwards or forwards, the voltage at the center pin (the output) varies between 0 and 5 volts.

The second is the purpose of the sketch, which is actually very simple – the voltage at analog pin A0 (the potentiometer's output) is measured and when it reaches a threshold value (400 in the sketch), the LED connected to pin 13 is turned on. When it is below the threshold, the LED is off.

The first block of code contains the variables:

```
const int analogPin = A0;    // pin that the sensor is attached to
const int ledPin = 13;       // pin that the LED is attached to
const int threshold = 400;   // an arbitrary threshold level that's in
                                the range of the analog input
```

Here we have three variables: The first two, "analogPin" and "ledPin" are the pins the potentiometer and the LED are connected to respectively. The third, "threshold", is an arbitrary number chosen as the condition at which the LED is turned on or off.

You will notice the "const" preceding each of these variables. This is a keyword for constant. The purpose of these constants is to make sure the value of the variables never change, i.e. the LED is always connected to pin 13, the threshold is always 400, etc.

Next is the setup() function.

```
void setup() {
   // initialize the LED pin as an output:
   pinMode(ledPin, OUTPUT);
   // initialize serial communications:
   Serial.begin(9600);
}
```

The first function, "pinMode", sets the pin the LED is connected to as an output. The second, "Serial.begin()" initializes the serial monitor at a speed of 9600 bauds.

Following setup() is the loop() function. As with most sketches, this is where the majority of the sketch's code is found.

A potentiometer is basically a three-terminal resistor with a sliding or rotating contact that acts as a voltage divider.

Constants provide a kind of safety net that protect your sketch from the effects of unintended changes in values.

...cont'd

```
void loop() {
    // read the value of the potentiometer:
    int analogValue = analogRead(analogPin);

    // if the analog value is high enough, turn on the LED:
    if (analogValue > threshold) {
    digitalWrite(ledPin, HIGH);
    }

    else {
      digitalWrite(ledPin,LOW);
    }
```

Hot tip

In a nutshell, the loop code is saying if A is true, do B. Otherwise, do C.

160

At the top is a function called "analogRead()". This reads the value at pin A0 and assigns it to an integer variable called "analogValue". As the voltage at A0 is the output from the potentiometer, the line of code is checking the level of voltage.

Next, the value at pin A0 is compared with the threshold value of 400. This is done with an if statement, which states that if the value at A0 is greater than the threshold value, i.e. is HIGH, the LED will be turned on.

The final piece of code is an else statement. This states that if the condition in the if statement is not met, i.e. the threshold at pin A0 is not greater than 400, or LOW, the LED is turned off.

From this you can see that the else statement allows you to specify what will happen if the condition in an if statement is not met.

The final piece of code is:

Hot tip

This sketch also shows you a comparison operator in action; in this case the "greater than" operator >.

```
    // print the analog value:
    Serial.println(analogValue);
    delay(1);        // delay in between reads for stability
}
```

This employs the println() function to send the sketch's data to the serial monitor from where it can be viewed. This enables the performance of the sketch to be precisely monitored.

The delay() function slows things down so the data can be read more easily, in this case by 1 millisecond.

The ForLoopIteration Sketch

There are a number of functions that you can use when writing your Arduino sketches. Some are more useful than others and one that definitely falls into this category is the "for loop" function.

This basically lets you specify exactly the number of times an action is to be run. Not only does this save the programmer from having to write out the code for each instance, it also makes the code shorter and thus easier to read.

The Circuit

The circuit for this sketch will require the following parts:

- Six 220 ohm resistors
- Six LEDs
- Breadboard
- Jumpers

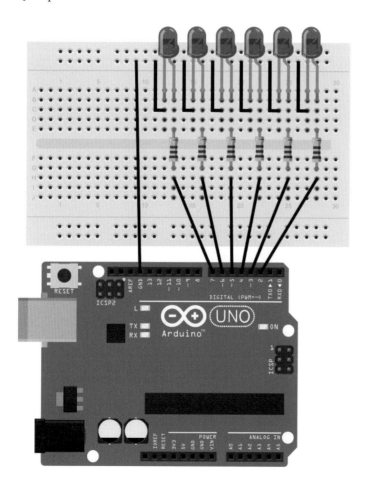

Hot tip

It is not necessary to lay out the parts exactly as shown here. Use your own layout but make sure the long leg of each LED is connected to a digital pin via a resistor, and that the other legs are connected to the common power rail.

...cont'd

Take six LEDs and connect them as shown. The long leg of each LED is connected to a digital pin (use pins 2 to 7) via a 220 ohm resistor, and the short legs are connected to the breadboard's common power rail. Finally, connect the common power rail to a GND pin on the Arduino board.

Connect the Arduino to your computer with the USB cable, open the IDE and go to **File > Examples > 05.Control > ForLoopIteration**. Click **Upload** and wait for the sketch to load.

You will now see the LED's turn on and off sequentially.

In the editor will be the following code:

```
int timer = 100;   // The higher the number, the slower the timing

void setup() {
  // use a for loop to initialize each pin as an output:
  for (int thisPin = 2; thisPin < 8; thisPin++)  {
  pinMode(thisPin, OUTPUT);
  }
}

void loop() {
  // loop from the lowest pin to the highest:
  for (int thisPin = 2; thisPin < 8; thisPin++) {
  // turn the pin on:
  digitalWrite(thisPin, HIGH);
  delay(timer);
  // turn the pin off:
  digitalWrite(thisPin, LOW);
  }

  // loop from the highest pin to the lowest:
  for (int thisPin = 7; thisPin >= 2; thisPin--) {
  // turn the pin on:
  digitalWrite(thisPin, HIGH);
  delay(timer);
  // turn the pin off:
  digitalWrite(thisPin, LOW);
  }
}
```

Hot tip

Note that in the sketch code on the right, we have not included the comment block at the beginning of the sketch.

...cont'd

Sketch Analysis

The first line of code is a variable:

int timer = 100;   // The higher the number, the slower the timing

This is an integer and it determines the time it takes the LEDs to turn on and off. As the comment says, the higher the number the slower the rate.

Next, we have the setup() function:

```
void setup() {
   // use a for loop to initialize each pin as an output:
   for (int thisPin = 2; thisPin < 8; thisPin++) {
   pinMode(thisPin, OUTPUT);
   }
}
```

This contains a For loop. Inside the brackets there are three separate statements: The first, "int thisPin = 2", initializes the pin the first LED is connected to, in this case pin 2.

The second statement, "thisPin < 8", determines whether the loop will run. If the condition is TRUE, i.e. 2 is less than 8, which it is, the code in the loop's braces will run. If it isn't, the statement will not be executed and the sketch moves on to the next line of code.

The final statement, "thisPin++", provides us with a way of incrementation. The ++ means add 1 to the value of thisPin.

To sum up the above then, if the condition in the loop is met, the code in the braces will be executed and the thisPin variable is incremented by a value of 1. The loop then runs again and assuming the condition is still met, the code is executed again and the thisPin variable increases again. This sequence continues until the thisPin variable has a value of 8. At this point the condition will now be FALSE, and thus the loop stops running.

Each time the thisPin variable is incremented by 1, the value of the "pinMode()" variable increases accordingly. Therefore, pin 2 through to pin 7 are successively set as OUTPUTS.

After the setup() function, we have the loop() function:

When using ++ and --to increment or decrement a value, by default the figure used will be 1 – it does not need to be specified. All other values do need to be specified, however. For example: thisPin++6 or thisPin--9.

This sketch shows the advantage offered by the For loop. Without it, we would have to write a separate instance of the pinMode variable for each of the LEDs. With it, we need just one.

...cont'd

```
void loop() {
  // loop from the lowest pin to the highest:
  for (int thisPin = 2; thisPin < 8; thisPin++) {
    // turn the pin on:
    digitalWrite(thisPin, HIGH);
    delay(timer);
    // turn the pin off:
    digitalWrite(thisPin, LOW);
  }
```

The "digitalWrite(thisPin, HIGH)" function has two arguments – the pin number and the output (HIGH or LOW). The HIGH constant applies 5 volts to thisPin thus turning on the LED connected to the pin. The pin in question is pin 2 as specified by the thisPin variable.

Next, a delay is introduced so the LED remains lit for a short period. This is achieved with the "delay(timer)" function, which stops things for 100 milliseconds. Then the "digitalWrite(thisPin, LOW)" function writes a value of LOW to pin 2, removing the voltage and turning the LED off.

The loop now starts again, incrementing the counter variable to 3 thus turning on the LED connected to pin 3. The loop continues to run until the last LED (at pin 7) has been turned on.

Now, we need to reverse the process to get back to pin 2. This is achieved with the final piece of code, which is another loop.

```
  // loop from the highest pin to the lowest:
  for (int thisPin = 7; thisPin >= 2; thisPin--) {
    // turn the pin on:
    digitalWrite(thisPin, HIGH);
    delay(timer);
    // turn the pin off:
    digitalWrite(thisPin, LOW);
  }
```

This works in the opposite way to the first loop. It starts with the thisPin variable set to pin 7. The test condition is now "is this pin greater to or equal to 2". The thisPin-- statement subtracts 1 from the value of thisPin each time the loop runs. When the value is back to 2, the test condition is not met and the loop stops. Now back at the top of the first loop, it all runs again.

Beware

The Delay() function is only recommended for use with short delays as it can bring a program to a halt. Alternative methods include:
- millis()
- micros()
- delayMicroseconds()

10 Troubleshooting & Debugging

Any number of things can go wrong with Arduino projects. We look at some common issues and methods of resolving them.

Before You Start!

Experienced Arduino users know that the next problem is always just around the corner. The more complex the project, the more certain this is. It goes without saying, then, that beginners are going to struggle, even with the most basic projects.

Knowledge

No matter how experienced you are, problems are also guaranteed by the number of different skills required by many projects – electronics, programming, computing, mechanics, woodwork, metalwork, etc. These all throw more complexities and issues into the mix.

Clearly, the more understanding you have of what you are doing, and how the various parts of a project interact with each other, the more likely it is that you will be able to resolve issues as and when they arise. Knowledge, therefore, is one of the most important weapons in your armory. Learn as much as you can, particularly with regard to electronics, computing and programming. Without a good grasp of these three skills, you will struggle to get anywhere with Arduino.

Technique

Also important is how you go about troubleshooting. Many people take the bull-in-a-china-shop approach by diving straight in and fiddling with this, that and the other. Usually, they end up making matters even worse.

A much more sensible approach, one that is almost guaranteed to yield better results, is to first consider what is happening (or not as the case may be) in a logical manner. Eliminate parts of the project that can have no bearing on the issue – if you can rule out 50% of it, then you are 50% more likely to find the cause of the problem. Isolate as much as you can.

Investigate any changes you have just made to something that was previously working. If possible, undo the changes and then check if the problem has been resolved.

Make a list of everything you think could cause the problem in order of likelihood and then work your way through the list.

If you still can't get it working, try posting a question in one of the Arduino online forums, such as **http://forum.arduino.cc** These can be a very useful source of information.

Hot tip

See the inside back cover for skills covered in the In Easy Steps series.

Hot tip

Very often, considering what is working rather than what isn't, will point you in the right direction.

Hot tip

Online forums are a great place to see if other people have had the same problem, or to ask for advice. There is a large Arduino community online and you are sure to get assistance.

Hardware

Faults with your projects can be either hardware or software related. Hardware issues can be checked out as explained on this page. Software issues are dealt with in the rest of the chapter.

The Arduino Board

Depending on the type of problem you are experiencing, making sure the Arduino board is functioning and correctly connected to the computer may be one of your first moves.

Work through the following check list:

- Make sure the computer is switched on (you'd be amazed by how many people neglect this most basic of steps).

- Connect the Arduino to the computer with the USB cable.

- Check the PWR LED on the board. This should be glowing a nice shade of green. If it is, it shows the Arduino is connected to the computer and is powered up.

  If the PWR LED is off, or faint, first check the USB cable is securely connected to both the computer and the Arduino. Then try connecting the cable to a different USB port; it's not unknown for these to become faulty. Finally, replace the cable with a new one.

- If you are using an external power adapter rather than a computer to power the Arduino, ensure that it is working correctly – you can do this by connecting it to a different device. Also, make sure the connecting cable is good, and that it is connected to the correct pins on the Arduino board.

External Hardware

External hardware includes circuits connected to the Arduino board, sensors, motors, etc. These all need power to function and this is the very first thing to check. If power is missing, very often the problem will turn out to be nothing more than a bad or missing connection.

Once you've eliminated power supplies and connections, isolate the problem by substitution where possible. Many parts, such as sensors and motors, can be checked in this way.

Faultfinding on circuit boards requires test equipment, such as a multimeter (and the knowledge to use it). See pages 72-73.

Bad and incorrect cable connections are a common cause of hardware issues. Be sure to always check these out first.

Substituting parts with ones known to be good is an excellent way of fault finding.

Set Up Issues

On pages 34-36, we explained how to set up Arduino with a computer. You may experience problems when doing this and, to a large degree, these will depend on what operating system your computer is running.

Windows

Windows users generally find it relatively straightforward to set up Arduino. This is particularly so if you are using Windows 8 – this operating system does it all for you automatically.

However, if you are running Windows 7, Windows Vista or Windows XP, you are likely to experience problems. This is because these operating systems have can have trouble locating and installing the Arduino driver. This manifests itself with the following error message (or similar):

The solution is to install the driver manually. Do this as follows:

1. Go to the Start Menu and click **Control Panel**. Then open the Device Manager. This lists all the hardware installed on your system, including the Arduino board

2. Look down the list until you see an Arduino entry (if you don't see one, it may be listed under Ports). This should have an exclamation mark next to it, which indicates it has not been installed correctly

3. Right-click on the Arduino entry and select **Update Driver Software** in the options menu. Then click **Browse My Computer for Driver Software**

4 Click **Browse** and go to C:\Program Files (x86)\Arduino\ Drivers. Here you will see a .inf file called Arduino. Select it and click **Next**. Windows will install the driver

Your Arduino should now be communicating with the computer.

Mac OS X
The Arduino setting up procedure for Mac OS X Lion, Mountain Lion, Leopard and Snow Leopard is the same and should be straightforward.

However, as with older Windows and Linux operating systems, setting up Arduino on older editions of Mac OS X can give rise to issues.

The Arduino software for Mac OS X is supplied in a zipped package and when the files are extracted, i.e. unzipped, corruption can occur. Sometimes you will get an error message so you'll know what the problem is but sometimes you won't.

Whichever, try unzipping the package with a different zip program. There are many of these available – two good examples being 7-Zip and Winrar.

Another issue that can occur is having an outdated version of Java on the computer. This should bring up an error message to this effect; if so simply download the most recent version from the Java website.

If you get a Link (dyld) error message, the solution is to upgrade to a more recent edition of Mac OS X. Old editions have incompatible system libraries.

Linux
As we noted on page 36, setting up Arduino on older versions of Linux can be a complicated issue, which we don't have room to adequately address here.

However, there are a number on online resources where you can find help, such as **http://www.linux.org**

An alternative method of finding the driver is to enter Arduino in the Windows Search utility. You will find this on the Start menu.

169

If you are considering using Arduino with Linux, the best advice we can give is to use the latest version. This will eliminate all the many issues that occur with older versions.

Syntax Errors

The Arduino is a computer, albeit a very simple one. Like all computers it doesn't think for itself, rather it simply reacts to the instructions it is given. These instructions (the sketch code) must be written in a way that the computer recognizes otherwise it just ignores them and does nothing.

Syntax is the term used to describe a set of rules that govern the way program code is written.

The rules that govern the way a program's code is written is called syntax and it is something you must learn in order to program your Arduino. If you don't, you will find it impossible to write a sketch that the Arduino's verification process, or compiler, will accept.

Getting syntax wrong is probably the most common mistake made by beginners and it always results in the same thing – the verification process simply stops when it sees the mistake.

Hot tip

170

Common syntax errors include forgetting the semi-colon at the end of statements, omitting opening or closing brackets and braces, and misspelling commands.

Fortunately, the compiler tells you where the mistake is and also its nature. Consider the example below:

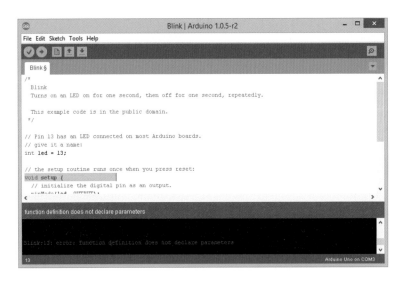

Beware

The description of the mistake is often cryptic and may not mean anything to the beginner.

The compiler has spotted that the brackets are missing at the end of the void setup function. It should read "void setup()". It has helpfully highlighted the line for you, and in the message window at the bottom it has stated that a parameter has not been declared for the function on line 13.

This is all the information you need to correct the mistake.

Serial Monitor

The serial monitor is a feature of the Arduino IDE that enables you to communicate with the serial port on the Arduino board. With it, you can send and receive data. Received data can be viewed in a window on the computer as we saw in Chapter 9.

This ability to send and receive data, and particularly to view received data, makes the serial monitor a useful debugging tool. With it, you can analyze code for errors and quickly correct them.

Before you can use it, you need to connect the Arduino to your computer with the USB cable. Then you have to open the serial monitor's window. This can be done in three ways:

 Open the Arduino IDE and from the menu bar at the top, select **Tools > Serial Monitor**

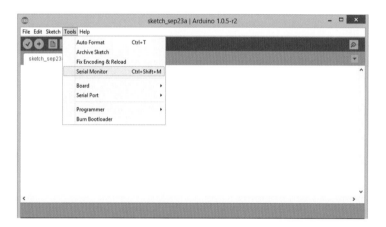

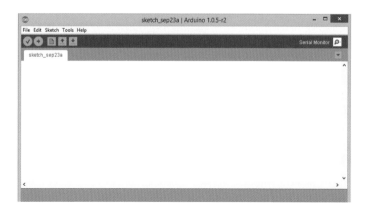 Click the **Serial Monitor** tab on the toolbar

The serial monitor is used to debug Arduino sketches and to view data sent by running sketches.

The serial monitor is effectively a separate terminal.

...cont'd

 Press the Ctrl + Shift + M key combination on your keyboard

Whichever way you do it, the serial monitor window will appear as shown below:

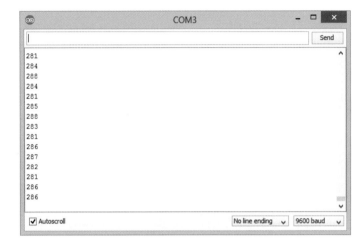

If it doesn't, the problem may be that the Arduino is using the wrong port. Open the Device Manager in Windows (as explained on page 168), and click the arrow to the left of Ports (COM & LPT). This will tell you which port the operating system has assigned to your Arduino.

Go back to the Arduino IDE and from the menu bar at the top, select **Tools > Serial Port**.

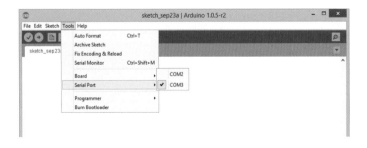

Ensure the port selected is the one assigned to the Arduino board in the Device Manager. You should now be able to open the serial monitor window.

Sending text to the Arduino via the serial monitor is done by entering the text in the box at the top of the window and then clicking the **Send** button.

Note that the serial monitor will not be able to send or receive data if you haven't included the requisite code in the sketch, i.e. told it to do so – see page 174.

At the bottom of the serial monitor window, you will see three settings:

- **Baud Rate** – this is the speed at which the serial monitor sends and receives data to and from the Arduino. The default setting is 9600 baud.

- **Line Ending** – this gives you four options with which you can set the type of end-of-line marking the serial monitor sends. Apart from New line, you can choose from No line ending, Carriage return and Both NL & CR.

  Carriage return and New line are the ASCII characters sent when you press Enter on the keyboard. Carriage return indicates the cursor will be returned to the beginning of the line, while New line indicates the cursor moves to the beginning of a new line.

- **Autoscroll** – if you want the last line in a sketch to always be displayed, check this box. Otherwise, you will have to scroll through the code manually.

Useful though the serial monitor provided with your Arduino is, there are alternatives. These include:

Processing – a free application for Windows, Mac and Linux

CoolTerm – a freeware application for Windows, Mac and Linux

puTTy – an open-source application for Windows and Linux

CuteCom – an open-source application that runs on Linux

You must tell the serial monitor that you want it to send and receive data by including code in the sketch.

When communicating with the Arduino, you aren't restricted to the serial monitor. There are a number of third-party programs you can use.

Debugging

Before the serial monitor can be used to troubleshoot a sketch, or debugging as it is known, you need to activate it and then tell it what you want it to do. This means inserting two pieces of code into the sketch being analyzed.

The first piece of code activates the serial monitor and looks like:

Serial.begin();

The Serial.begin() function is just one of a number of functions provided by the Arduino's built-in serial library. Inside the brackets, you need to enter the desired communication speed, i.e. baud rate. 9600 is the standard rate and equates to approximately 1,000 characters per second. The syntax is thus:

Serial.begin(9600);

Whatever baud rate you enter between the brackets, make sure the same rate is selected in the serial monitor – if the rates are different, you will get nothing but gibberish in the monitor's window.

The second piece of code is the important bit with regard to debugging. This is:

Serial.println()

The Serial.println() function tells the serial monitor to display text in the monitor window (without it, the serial monitor will run but its window will be blank).

If you specify a statement between the brackets – the output from a sensor, for example – the Serial.println() function will display that value in the monitor window. This enables you to see at a glance if the value is correct. If the value isn't displayed, you know there is a problem and its approximate location.

For debugging purposes, you can use the Serial.println() function as follows: If you have no idea where the problem code is, you can simply place numerous instances of Serial.println() throughout the sketch and then as the sketch is run, literally follow along in the monitor window.

Or, if you suspect a certain section or line of code, you can troubleshoot in a more selective manner.

Don't forget

Your sketch must call the Serial.begin() function before it can use the serial monitor. The function is usually placed inside the setup() function.

Beware

If the selected baud rate does not match the value set in your sketch, characters on the serial monitor will be unreadable.

Hot tip

The Arduino serial library contains a number of functions that can help you to debug code.

11 Arduino Projects

A good way of discovering just what can be done with Arduino is to look at projects built by other people. We show a selection in this chapter. You can, of course, find many more online.

Introduction

In the preceding chapters we have attempted to give you a grounding on the skills and knowledge you need to build electronic circuits and put your Arduino to good use. We are now going to showcase some existing projects that are built around the Arduino. Not only do these show you what can be done with it, they demonstrate the range of skills needed to build complex Arduino projects.

GSM Security Alarm System

The purpose of this project is to provide a long-range security alarm system that can be used in the home (or anywhere else for that matter).

It consists of an Arduino Uno, a standard SIM900A based GSM/GPRS modem, and an intrusion-detector unit such as an infrared proximity sensor or a light fence. The system is powered by a 12 volt DC power supply unit or battery.

Hot tip

Most Arduino projects require a number of skills. These can include, programming, electronic circuit building, sensor interfacing, enclosure construction and more.

When the system is triggered by an intrusion attempt, a warning SMS text message is sent to a specified cell phone. The system also provides a "call-alert" feature, which initiates a telephone call when it is activated. This makes a "missed call" alert.

Thanks to www.electroschematics.com for permission to feature the GSM Security Alarm System in this book.

LED Cube

Incorporating no less than 512 LEDs, plus a large number of shift register integrated circuits (ICs), resistors, transistors, capacitors, and a huge amount of wiring, the LED Cube serves no useful purpose whatsoever.

It is, however, a fascinating object that offers an almost limitless number of light permutations that will prove highly entertaining – kids, for example, will love the pyrotechnic light show it provides.

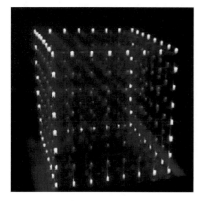 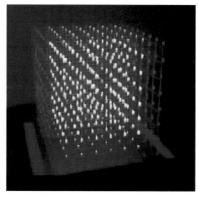

The LED Cube consists of eight layers of LEDs spaced equidistantly and wired together with .22mm wire. Due to the wide spacing, the LEDs on all eight layers can be seen simultaneously giving the cube a third dimension.

This is a classic Arduino project involving, as it does, a number of different skills. Not only do you need to know how to program the Arduino, you will also need to know how to design and build electronic circuits.

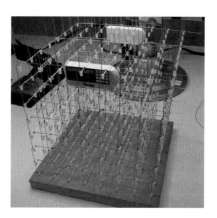

Furthermore, each layer has to be exactly the same as the others and this requires a very precise construction technique.

Thanks to Caleb Picou for permission to feature the LED Cube project in this book.

Hot tip

If you decide to have a go at this project, remember that you will need hundreds of LEDs. These are readily available online at very low prices. Their quality, however, may be equally low. Unless you are happy to be constantly replacing faulty LEDs, we suggest you get good quality ones.

177

Hot tip

The physical aspect of this project, i.e. building and stacking the layers into a robust construction is not easy. Building a much smaller 2 x 2 x 2 cube first to figure out the best way of doing it is recommended.

Skube

Skube is a project that originated at the Copenhagen Institute of Interaction Design (CIID) as part of the Tangible User Interface module. It demonstrates just how useful Arduino can be for prototyping and development.

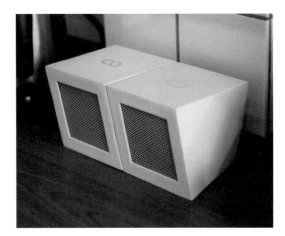

Two Skubes connected together

Skube is a portable wireless Internet radio. It connects to your Last.fm account to load your music, and uses Spotify to find and play the tracks.

The motivation for the development of Skube was the realization that with the trend towards listening to digital music online, current portable music players are not adapted for this environment and are thus unsuitable for it.

Add to this the fact that sharing music in communal spaces is neither convenient nor easy, especially when people have such different tastes in music.

The result is a music player that enables you to interact with digital music services such as Spotify without the need for a computing device – everything is controlled from the Skube.

Each Skube provides two modes – Playlist and Discovery. These are selected by simply tapping on the top of the Skube. Playlist plays the tracks on your Skube, while Discovery looks for tracks similar to the ones already on your Skube. This lets you easily find new music that is likely to appeal to your particular tastes.

When two Skubes are connected, their playlists merge into one. If you like a track from your friend's Skube, simply click the heart button at the back of yours to add it to your own playlist.

When more than one Skube is connected together, they act as a single player that shuffles between all the playlists. This enables a number of Skubes to be controlled from one device.

The interface is designed to be intuitive and simple. Flipping the Skube changes modes, tapping will play or skip songs, and flipping a Skube on to its front face will turn it off.

The heart of the system is Arduino and all Skubes have an Arduino board inside. They also have an XBee wireless shield, which provides wireless end-point connectivity to devices. This enables individual Skubes to communicate with each other.

Skubes contain a number of sensors that respond to the various actions needed to control them, such as tapping, flipping, etc. The inputs from these sensors are fed to the Arduino, which interprets them and sends the appropriate signals to the rest of the system.

Hot tip

XBee shields use the IEEE 802.15.4 networking protocol for fast point-to-multipoint or peer-to-peer networking.

The devices also contain an FM radio shield. The purpose of this is to play the music.

The Xbee wireless shields mentioned above are also used to communicate with, and coordinate, Skubes with a computer.

This is done with the aid of a visual programming language called Max, which has been developed for music and multimedia.

Two well known music services, Spotify and Last. fm, both provide a resource known as an Application Programming Interface (API). Max takes data from these API's and uses it to provide Skube's Playlist and Discovery features.

Hot tip

Spotify's Web API lets applications fetch data about artists, albums, and tracks directly from the Spotify catalogue. The API also provides access to user-related data such as playlists and music saved in a "Your Music" library.

The Skube project involves a number of elements including programming, circuit construction, wireless communication, data sharing with external services, and enclosure construction.

Thanks to Andrew Spitz for permission to feature the LED Cube project in this book.

Lawnbot400

Some people enjoy mowing grass; others hate it! If you are in the latter category why not do something about it? If you're wondering what exactly, this project may be just what you are looking for.

The Lawnbot400 will prove to be a good test of your mechanical and electronic skills.

The Lawnbot400 is a remotely controlled lawn mower that consists of the following:

- A push lawn mower with the wheels and handlebars removed

- Two 12 volt batteries to provide 24 volts of power

- A sturdy metal frame and tray to support the mower and the batteries

- Two electric motors to move the mower

- A remote control transmitter and receiver with which to control the mower

- Electronics, including an Arduino board

The metal frame and wheels are a major part of the project. Construction requires a good grasp of mechanical skills not to mention the requisite tools. There are no hard and fast rules here – ingenuity will be your best friend.

Once the frame is done, you will need to mount the motors. Any suitable motor can be used – the ones in the Lawnbot400 were taken from a wheelchair. Whichever motors you use, their position will have to be adjustable to allow the tension of the drive chain to be adjusted. This will require some form of mounting plate to be fabricated.

The speed of the lawnbot is controlled by a purpose-built motor controller. This supplies a variable voltage to the motors. The controller is itself controlled by a pulse width modulation (PWM) signal sent from an Arduino board. The motor controller converts the 0 - 5 volt PWM values into 0 - 24 volt direct current voltage that is sent to the motors.

The next stage is fitting the mower to the frame. Again, this will be a test of your mechanical skills and ingenuity.

The final part of the mechanical construction is fitting the batteries. These are heavy items, so fitting them behind the rear wheels greatly improves control of the machine as they counter the weight of the mower in front.

Steering the Lawnbot is straightforward. Move the left control stick up, and the left wheel moves forward. Move the right control stick back, and the right wheel moves backward. Move both sticks forward and you go straight ahead. This is called "tank steering," and it gives the Lawnbot400 a zero turn radius.

The Lawnbot has the potential to be a very dangerous piece of machinery. For this reason it is essential that the frame and the mower attachement bars are built to a high standard.

To make sure the user doesn't lose control of the Lawnbot, a fail-safe device has been built in. This consists of another Arduino that controls a 60 amp power relay. This disconnects the power to the motor controller unless it is receiving a "good" signal from the Arduino.

Finally, there is a kill-switch on the transmitter that cuts the power to the motors, thus instantly disabling the machine should it be necessary to do so.

Thanks to John-David Warren for permission to feature the Lawnbot400 project in this book.

Safety is an important consideration in this project.

The Lawnbot400 as it stands has the potential for improvement. For example, it could be completely automated by incorporating a GPS system and sensors. You could also connect an electric motor to the mower drive shaft to automatically charge the batteries.

Baker Tweet

The best time to buy pastries and other baked goods is when they are fresh out of the oven. With this indisputable fact in mind, the team at POKE's London office decided to ensure that their favorite bakery (which just happened to be across the road) had a way of instantly updating them with regard to this important issue.

Baker Tweet has helped put the Albion Bakery on the map.

To this end, they have designed and built a wall-mounted Wi-Fi device with a large, dough-friendly dial. It has a bespoke Arduino circuit board and hygienic, wipe-clean controls that can announce the arrival of freshly baked goods on Twitter as soon as they are ready to eat.

All the baker has to do is turn the dial to the required option and press the button below. All of the baker's Twitter followers get an alert telling them what's just come out the oven.

The Baker Tweet device has been installed at the Albion Cafe on Boundary Street in Shoreditch, London, and has proved to be very good for business. Other bakeries would no doubt benefit in the same way.

It's not just bakeries that can benefit, however. Baker Tweet enables any business to communicate with its customers in real-time via Twitter.

There are any number of potential uses for a Wi-Fi connected device that can automatically update a social network such as Twitter.

By logging on to a web interface (which is supplied as part of the package), it is possible to send information regarding offers, pricing and stock from a wall-mounted device that is sturdy enough to withstand the rigours of a busy kitchen, and is much simpler to operate than a laptop or a mobile.

The Baker Tweet is constructed of the following parts:

- Arduino Duemilanove board

- Arduino Ethernet Shield

- Ladyada Proto Shield

- Linksys Wi-Fi Adapter

- Assorted electronics, wiring, knob, switch, casing, etc.

The Baker Tweet device has been built with an Arduino Duemilanove. However, it can just as easily be built with an Arduino Uno.

One of the features of Baker Tweet is that all relevant information is stored on a website – this is built around the Django content management system for ease of use.

This means that an owner can sign into their account on **bakertweet.com** and edit the details of their products, e.g. the number of items, how they want them labeled on the device, etc.

Then they update the device by simply turning the dial to "Update Items List," and pressing the button.

*Thanks to the POKE team at **www.pokelondon.com** for permission to feature their Baker Tweet project.*

Django is a free and open source content management system that is used to publish content on the Internet.

Tree Climbing Robot

Many Arduino projects are simply exercises to see if an idea can be put into practice – they may serve no practical purpose but they do present a challenge. The Tree Climbing Robot is a typical example of this.

All the device does is climb up and down a tree trunk as shown in the image below. However, getting it to do this is far from straightforward and will exercise your design, electronic and mechanical skills.

Hot tip

The robot has four pairs of legs. Each pair is driven directly by its own motor.

This project needs a good deal of initial planning and design work. There are any number of suitable software programs for this purpose but the one used here was Sketchup.

The brief was to design the following:

A robot consisting of two segments, joined by a spine which could be extended or retracted. Both segments to have four legs with very sharp points as feet.

Hot tip

The climbing sequence is somewhat similar to the way an inchworm climbs.

To climb, the legs on the top segment pinch together and the feet dig into the bark, securing the robot. Then the spine retracts, pulling up the bottom segment. The legs on the bottom segment then grip the tree, and the top segment releases its grip. Finally, the spine extends, pushing the top segment upwards. The process then repeats.

The main parts of the robot are:

- Arduino Uno

- Motor controller

- Batteries and regulator for power

- Four motors

- Hardware to construct the robot

The motor controller is purpose built and is based on the L298HN Dual Full Bridge chip. It has the capability to bidirectionally control four DC motors, one for each of the legs.

The robot's power is supplied by two different sources. The Arduino and the motor controller logic circuitry are powered by a 9 volt battery, while the motors are powered by a 12 volt Li-Ion battery pack.

The robot has four pairs of legs, each pair controlled by one motor. These are made from lengths of aluminium bar. The main frame, which holds everything together is also constructed from aluminium.

To move up and down, the robot extends and contracts by spinning a threaded rod that is fixed to the top segment. When the rod is spun clockwise, the two segments are pulled together, and they are pushed apart when it spins counter clockwise. The threaded rod is powered by a 12 volt motor.

Moving on to the electronics, rotation sensors are key to the operation of the robot. It has one rotation sensor per motor, so the robot knows the exact position of each leg at all times, thus enabling them to be precisely controlled.

The wiring includes connecting up the motors, motor controller, the sensors, the regulator and switches.

With regard to the programming, this was done by writing a separate function for each basic action of the robot – see margin note.

Thanks to Ben Katz for permission to feature the Tree Climbing Robot project in this book.

Because a standard DC motor has been used rather than a servo or a stepper to spin the threaded rod that forms the spine, the robot can not know the degree of extension of the spine. Therefore, limit switches are used to prevent it from extending or contracting too much.

185

The functions are: closeTop, closeBottom, openTop, openBottom, Lift, Push. By combining these functions in the proper order, the robot can be made to ascend and descend.

8BitBox

The 8BitBox is a portable Bluetooth box that can be controlled via any Bluetooth enabled Android device, such as a smartphone or tablet. With it, you can remotely play and control music. It can also be used as a cool night light courtesy of the built-in LED.

Hot tip

This project will give you a good understanding of how to remotely control an Arduino project via Bluetooth.

Hot tip

If you use an RGB LED, you will be able to adjust the color of the LED.

The housing is an interesting project in itself as it is constructed with a 3D printer (although you can use your own method of construction). 3D printing is something that will be new to most people.

Vellum paper is fixed to the inside of all four sides to diffuse the light from the LED and provide a subdued lighting effect.

The electronics consist of an Arduino board that acts as the controller, an Arduino proto shield on which to mount the components, an RGB LED to provide the light, a piezo buzzer that plays the music, an EZ-Link module that provides the Bluetooth connection and a LiPo charging circuit to keep the battery fully charged.

Once assembled, everything is placed inside the box. Press the on/off button situated on the lid to operate the device.

*Thanks to Adafruit at **www.adafruit.com** for permission to feature the 8BitBox project.*